The Glass of Wine

The Glass of Wine
The Science, Technology, and Art of Glassware for Transporting and Enjoying Wine

James F. Shackelford
Penelope L. Shackelford

WILEY

This edition first published 2018

Registered Office
John Wiley & Sons, Inc., 111 River Street, Hoboken, NJ 07030, USA

Editorial Office
111 River Street, Hoboken, NJ 07030, USA

For details of our global editorial offices, customer services, and more information about Wiley products visit us at www.wiley.com.

Wiley also publishes its books in a variety of electronic formats and by print-on-demand. Some content that appears in standard print versions of this book may not be available in other formats.

Library of Congress Cataloging-in-Publication Data

Names: Shackelford, James F., author. | Shackelford, Penelope L., author.
Title: The glass of wine : the science, technology, and art of glassware for transporting and enjoying wine / By James F. Shackelford, Penelope L. Shackelford.
Description: Hoboken, NJ : John Wiley & Sons, 2017. | Includes bibliographical references and index. |
Identifiers: LCCN 2017041012 (print) | LCCN 2017046174 (ebook) | ISBN 9781119223467 (pdf) | ISBN 9781119223450 (epub) | ISBN 9781119223436 (cloth)
Subjects: LCSH: Glassware. | Wine glasses. | Wine bottles.
Classification: LCC TP865 (ebook) | LCC TP865 .S425 (print) | DDC 642/.7–dc23 LC record available at https://lccn.loc.gov/2017041012

Cover image: Reproduced with kind permission of Riedel Glas Austria
Cover design by Wiley

Set in 10/12pt TimesLTStd-Roman by Thomson Digital, Noida, India

10 9 8 7 6 5 4 3 2 1

To

Scott, Megumi, Mia, and Toki

Contents

Preface

Glass confronts us at every turn. The windows in our homes, the vessels for drinking, the spectacles that improve our vision, the microscopic fibers that provide modern communication – all are made of glass. So common and taken for granted – too often we pay little attention to glass unless it breaks. In the title *The Glass of Wine*, the word glass is at the heart of the book. Today wine remains the one beverage for which glass is used so pervasively in its making, containing, transporting, and drinking.

From the perspective of Davis, California, the home of a world-renowned wine program at the University of California, Davis, and James' work in glass science, Penelope's in art, and their combined passion for touring the wine regions of the world, *The Glass of Wine* developed organically over four decades. Initially, we learned about the basic bottle shapes associated with the common varietals. Then, we also came to recognize which size and shape of stemware were used for the various varietals: a cabernet versus a chardonnay and on and on. Touring made us aware of the ever-present glass equipment used in the laboratories of wineries, and, during these travels, we regularly happened upon glass collections in museums and the private collections of wine makers. As the concept of a book fully devoted to the unique and ubiquitous role of glass in wine emerged, historical research followed and the discovery of just how intimately the worlds of glass and wine have been intertwined over so many centuries.

Countless books have been written on wine, regions, tours, and tastings, but unlike others *The Glass of Wine* focuses on the glass itself and records its importance to the development of wine over three millennia. We hope its readers will find the same fascination with this unique and profound role of glass in the world of wine as we have. Santé, Salute, Salud!

Acknowledgments

We wish to thank a number of people who have been most helpful in bringing this book to fruition. First and foremost we would like to thank the excellent staff of The American Ceramic Society, especially Greg Geiger and Eileen De Guire who helped connect us with their publishing partners at Wiley and especially there Michael Leventhal, Divya Narayanan, and Beryl Mesiadhas who guided us expertly through the final stages of manuscript preparation. Writing a book about the intersection of two major industries, wine and glass, could not be possible without the support of key components of both industries. You will find specific credits throughout the book for images made possible through the courtesy of numerous wineries, winemakers, fine dining establishments, and glass producers. These brief acknowledgments in fact do not do justice to what was often an extended and joyous interaction. The many images from the Six Senses Restaurant in the Duoro Valley of Portugal in Chapter 13 are noteworthy in that they archive a fascinating convergence of glass technology and wine appreciation in the use of Port Tongs by the skillful hands of Wine Director, Mario Claro.

Special thanks are due to the generosity of the Corning Museum of Glass (CMOG) and the associated Rakow Research Library in Corning, New York. The Library was exceptionally generous in time and consultation for James during the time he was attending an ACerS Glass and Optical Materials Division meeting in Corning in 2010 during the early research phase on the book. This was followed by the equally generous support of the CMOG staff in providing an incredible set of high-resolution images of many priceless pieces of glassware from their collection. In the same way, Riedel and Luigi Bormioli, leading suppliers of fine wine glassware, provided images of their production technologies and the stunningly beautiful products of those technologies.

As noted a number of times throughout this book, our exposure to the wine world has been enhanced by being near the wine program at UC Davis, the leading program in the United States and one of the premier ones in the world. We have been most fortunate to know Dr. Clare Hasler-Lewis, Executive Director of the Robert Mondavi Institute (RMI) for Wine and Food Science. RMI houses the Department of Viticulture and Enology whose faculty members have been generous with their time and knowledge over the years, with special appreciation to colleagues Roger Boulton and David Block who have had joint appointments in the Department of Chemical Engineering and Andrew Waterhouse who provided James and his camera open access to the laboratories around RMI as well as the LEED Platinum Teaching Winery. We are also indebted to Dr. Patrick

McGovern, from the University of Pennsylvania Museum of Archeology and Anthropology (the Penn Museum). Dr. McGovern is without peer in uncovering the history of wine and other fermented beverages. His generosity, along with that of his colleagues at the Penn Museum, was essential to our ability to view the history of winemaking through our lens of glass.

Discussing the science and technology of glass was greatly aided by the kindness of James' editor at Pearson, Holly Stark, who provided the permission to use numerous illustrations from his textbook, *Introduction to Materials Science for Engineers*. Visualizing the nature of glass on the atomic scale was aided by Professor Sabyasachi Sen of the Department of Materials Science and Engineering at UC Davis who provided numerous computer-generated images and identified an additional critical example from his colleague Professor Jincheng Du of the Department of Materials Science and Engineering, University of North Texas.

Some individuals not seen in these pages are nonetheless equally worthy of our thanks. Professor Lilian Davila of the University of California, Merced, is a friend to both of us while she continues the computational studies of glasses and related materials she began as a student here at UC Davis, including illustrations of the atomic arrangement in glass and ceramics. Lilian was the first to point out to us the double helix in common quartz, a structure not limited to our DNA. Also, within the glass science community, Professor Emeritus Linn Hobbs of MIT has shared not only a common interest with James in understanding the atomic structure of glass but also his deep knowledge of wine with us as we have crossed paths at various ACerS affiliated meetings around the globe while simultaneously instilling an appreciation of wine in the students of MIT for over three decades through his iconic wine course. Finally, as noted in the Preface and detailed further in Chapter 1, this book is a result of our more than four decades long touring of wineries around the world. No one has been more helpful in this journey than Tim Robertson of Robertson Wine Tours. His friendship and guidance in many of our tours over the past decade are among our most treasured memories.

About the Authors

James F. Shackelford has BS and MS degrees in Ceramic Engineering from the University of Washington and a PhD in Materials Science and Engineering from the University of California, Berkeley. Following a postdoctoral fellowship at McMaster University in Canada, he joined the University of California, Davis, where he is currently Distinguished Professor Emeritus in the Department of Materials Science and Engineering. He teaches and conducts research in the structural characterization and processing of materials, focusing on glasses and biomaterials. His current focus in teaching is doing so through online technologies. A member of The American Ceramic Society and ASM International, he was named a Fellow of The American Ceramic Society in 1992 and a Fellow of ASM International in 2011; he received the Outstanding Educator Award of The American Ceramic Society in 1996 and will receive the Albert Easton White Distinguished Teaching Award of ASM International in 2019. He has published well over 100 archived papers and books, including *Introduction to Materials Science for Engineers* now in its 8th Edition and that has been translated into Chinese, German, Italian, Japanese, Korean, Portuguese, and Spanish.

Penelope L. Shackelford has a BA degree in English and Philosophy from St. Joseph's College and an MA degree in Arts and Consciousness from the

Photographer: Daniela Wood.

John F. Kennedy University. After careers as a teacher and an art curator and gallery owner, she became a journalist, a role in which she has served as an arts writer for the *Davis Enterprise*, *Artweek*, and a variety of national arts publications. She served as Associate Editor for Arts for the journal *Multicultural Education*. In addition, she has written about winemaking in California, Washington, and New Mexico. She has traveled with James extensively exploring major winemaking regions domestically and around the world. She chronicles these travels in her blog: Travels with Penelope.

Chapter 1

The Perfect Material – for Wine

"Fill ev'ry glass, for wine inspires us,
And fires us
With courage, love and joy."

John Gay 1685 – 1732

The Beggar's Opera [1728], act II, sc. I, air 19

Glass is simultaneously one of the most ancient human-made materials and one of today's most sophisticated high-tech products. While born of some of the most common elements of the earth, glass with its transparent and poetic qualities has become central to the appreciation of wine. The histories of this beverage and the material that contains it are intimately intertwined. Over the centuries, improvements in winemaking have emerged, as glass vessels were developed to reveal the liquid within.

Today, after the wine is produced and then aged, generally in wooden barrels or stainless steel tanks, it is transferred to glass bottles for secure storage (and further aging) and transported to the consumer. The bottles are thus utilitarian but also aesthetic, serving as primary marketing tools. Their shape can signal the type of wine, and their appearance can be a contributing factor in the consumer's purchase. Once in our home, we pour the wine from bottle to drinking glass to enjoy in its own right or with food. After leaving the barrel or tank, most wine sees only a glass surface until its journey ends on our lips (Figure 1.1). The great Dutch painter Vermeer immortalized this final step in a seventeenth century painting (Figure 1.2).

In North America, 40 billion glass containers are produced for beverages per year representing 15% of the general beverage market. For many beverages,

The Glass of Wine: The Science, Technology, and Art of Glassware for Transporting and Enjoying Wine, First Edition. James F. Shackelford and Penelope L. Shackelford.

Figure 1.1 Glass is the essential material when wine tasting, in this case at the Wine Resort Leda d'Ittiri on the island of Sardinia. (Reproduced with kind permission of Annamaria Delitala, Wine Resort Leda d'Ittiri.)

plastic, aluminum, and cardboard containers are popular alternatives to glass. On the other hand, nearly 100% of the wine market continues to use glass containers (Figure 1.3). An additional benefit of these glass bottles is the ease with which they can be recycled.

Glass, however, is not without limitations. It is an inherently brittle material and more dense than plastic containers. Modestly careful handling can prevent bottle breakage, but reducing the weight of individual bottles, largely through

Figure 1.2 The Glass of Wine by Johannes Vermeer c. 1658–1660 (Reproduced with kind permission of Staatliche Museen Preussicher Kulturbesitz, Gemaldegalerie, Berlin/Art Resource Inc.)

Figure 1.3 Bottles are densely stacked while their wine ages at the Château de Beaucastel Winery in the Châteauneuf-du-Pape region of France. (Reproduced with kind permission of Château de Beaucastel.)

reduced wall thickness, is an important consideration to an increasingly energy conscious industry. The cost of transporting thousands of cases of wine is non-trivial even for relatively small wineries.

This book emerges from our research involving numerous wine tours and tastings as well as from our home in Davis, California, where the Department of Viticulture and Enology at the University of California, Davis has played a central role in the history of winemaking in California and especially the renaissance of the past 50 years, as California vintages have become peers of some of the world's great wines. The Department is, in fact, the only one within the University of California mandated by the State Legislature (in 1880 when there was but one UC campus in Berkeley). The Legislature realized even then that California had the potential to become one of the world's great producers of wine grapes and wine. The Department was formally established on the Davis campus in 1935 following the end of Prohibition. This establishment was typical of the Berkeley–Davis relationship. Judge Peter J. Shields who was also a prominent farmer in Northern California had drafted legislation early in the twentieth century to establish an agricultural college modeled after the Pennsylvania State College of his day. The large parcel of land appropriate for this college was located about one hour to the east near a convenient train station in the small town of Davis. The location was originally named the University Farm. These train trips out to Davis were regular journeys for students from the College of Agriculture at Berkeley. By 1960, over 3000 students were now doing their agricultural studies full time on the Davis campus. In that year, the Master Plan of Higher Education, the brainchild of the visionary President of the University of California Clark Kerr, established Davis and several other sites around the state as autonomous campuses.

Growth was rapid, and, by the time the authors arrived in 1973, as James assumed an assistant professorship in the Department of Mechanical Engineering teaching and doing research in the field of materials science and engineering, the campus had grown to over 15,000 students. His material of choice for academic research was generally glass, an interest that went back to an introduction to this material as an undergraduate in the highly specialized field of Ceramic Engineering at the University of Washington. A lecture there by an engineer from Corning Glass was transformative, with the slides showing technologically important materials that were stunningly beautiful (a triumph of industrial photography!). After a senior thesis studying the crystallization of a sodium silicate glass, he was on to UC Berkeley in 1967 to do a Ph.D. thesis on gas solubility in high-purity silica glass. (Yes, some small diameter gas atoms such as helium and neon are quite soluble in glass. It was a successful study that has led him to continue pursuing related research questions to this day.)

The authors' first (blind!) date was during this time and included a visit to the Italian Swiss Colony Winery in Asti, California at the time a major purveyor of better-than-average bulk *vin ordinaire* often sold in gallon jugs. A marriage eventually followed that blind date, and, 2 years later, we settled in Davis. While James settled into the faculty position at UC Davis, Penelope worked in the art world as a curator and gallery owner and eventually as an arts writer reviewing exhibitions on campus, home not only to a premier wine program but also to a leading department of art.

That tour of the Italian Swiss Colony Winery led to a favorite pastime of wine touring. Our journeys into the world of wine have taken us to vineyards throughout Europe, North and South America, and more recently Asia. Over these four decades living in Northern California, we have also had the privilege of watching the transformation of California's wine industry from its largely jug wine identity to the production of world-class vintages. We had arrived separately in this region at about the time Robert Mondavi was becoming a central force in that transformation. He had just opened his landmark Robert Mondavi Winery in 1966 in the heart of the Napa Valley (Figure 1.4). It was astonishing that, within a decade, the "Judgment of Paris" wine tasting would establish the best cabernet sauvignons and chardonnays[1] of the Napa Valley to be the full peers of their finest counterparts from France.

One of our early wine tours included visiting the Robert Mondavi Winery, and, in 1994, we had the privilege of meeting the man himself at a celebratory dinner at UC Davis to honor Professor of Art Wayne Thiebaud for his recent recognition with the National Medal of the Arts from President Clinton. Penelope covered the story of the Thiebaud dinner for the local newspaper. The dinner was also an opportunity for Chancellor Larry Vanderhoef to announce the vision of Robert and Margrit Mondavi to establish a center celebrating the

[1] We will follow the guide of Karen MacNeil, author of *The Wine Bible*, in the capitalization of wine names. For many varietals, such as cabernet sauvignon and chardonnay, the names are not capitalized as they are named after the grape variety. Only those wines named after a specific place, such as Champagne or Barolo, will be capitalized. Note that "variety" refers to a grape and "varietal" to the wine made from that grape variety.

Figure 1.4 Robert Mondavi, a leading pioneer in the production of fine wines in America, toasts with a glass of wine in front of the Robert Mondavi Winery in the Napa Valley of Northern California. (Reproduced with kind permission of Anne Siegel, Robert Mondavi Winery.)

interrelationship of wine, food, and art, a vision that within a decade would be realized with the Robert and Margrit Mondavi Center for the Performing Arts and the adjacent Robert Mondavi Institute (RMI) for Wine and Food Science on the Davis campus (Figure 1.5). Together these serve as a testament to a pioneering winemaker's vision of the role of fine wine and food along with the arts to enhance our quality of life. We certainly share his vision with the convergence of art and wine in our own lives through a lens of glass.

This was also about the time when the program in Materials Science and Engineering was transferred from the Department of Mechanical Engineering to the Department of Chemical Engineering. Now, James found himself a colleague of some who held joint appointments between Chemical Engineering and the Department of Viticulture and Enology. Seeing the world of grape growing and winemaking close at hand while continuing to do research on glass, along with our increasingly sophisticated touring of the wine country, including visits throughout California and north to the States of Washington and Oregon, created the nucleus of an idea – glass plays a unique role in wine culture, and it is a story worth telling.

In this book that relates the interweaving of wine with the glass that contains it, we will next move to the history of wine itself. In Chapter 2, the early history of winemaking and wine drinking does not involve glass, but generally ceramic pottery, animal skins, and metal chalices. The innate affection of humans for fermented

Figure 1.5 The Robert Mondavi Institute (RMI) for Wine and Food Science on the University of California, Davis campus is a fitting tribute to the California winemaking pioneer. Here a spectrum of wine bottles waits to be filled by students in the Teaching Winery at RMI. (Reproduced with kind permission of Andrew L. Waterhouse, Robert Mondavi Institute for Wine and Food Science.)

beverages predates their creativity in developing glass vessels by a few millennia. Wine and related "mixed" fermented beverages appeared in the early Neolithic period (7000–5000 BC) more or less simultaneously across all of Asia from the west (in the modern Middle East) to the east where the earliest fermented beverage discovery to date has been found in the center of modern China.

Chapter 3 provides us with a formal definition of glass, a first cousin of **ceramics** and distinct from the other major categories of engineered materials: metals, polymers (plastics), and composites. Chapter 3 also continues the story of wine storage and consumption, with the development of glass some time before 2000 BC, initially providing a fragile and expensive alternative. While winemaking had a few millennia head start on glassmaking, the two technologies became intimately intertwined in the seventeenth century, as inventions in England led to high-quality glass becoming widely available for bottles and drinking glasses. This period in time served as a fulcrum for the histories of both glass and wine, as increasingly transparent vessels revealed imperfections

Figure 1.6 Venetian glass such as this example from the late sixteenth to early seventeenth centuries was of such high quality that winemakers were inspired to make better, sediment-free wines. (Reproduced with kind permission of the Corning Museum of Glass.)

in the beverage. The convergence of the histories of glass and the wine it contained was most dramatic in the Champagne region of France as the transparent beauty of Venetian glass inspired winemakers to produce crystal clear and sediment-free sparkling wines (Figure 1.6).

Chapter 4 diverges from the general focus of this book on an important inorganic material (glass) and provides an overview of the related and equally important topic from organic chemistry: how wine is made. Humans have been making wine for a considerably longer time than they have been making glass. Contemporary winemaking and the precedent agriculture of grape growing, viticulture, involve a wide variety of materials, with glass playing a minor role. Nonetheless, we shall generally use the menu of materials available to winemakers as a kind of lens through which we view the art and science of winemaking.

Once harvested by hand or by machine, grapes can experience oak barrels (often from France, Hungary, or America) during vinification or, as an alternative, stainless steel tanks. These vessels are environments for a complex variety of reactions that ultimately can be simplified as the process of fermentation – the

Figure 1.7 Scientific glassware is essential in winery laboratories, in this case at the Robert Mondavi Institute for Wine and Food Science. (Reproduced with kind permission of Andrew L. Waterhouse, Robert Mondavi Institute for Wine and Food Science.)

conversion of sugars in the grape juice into alcohol. While Chapter 4 focuses on the organic chemistry of winemaking and materials other than glass in that service, we see in Chapter 5 that ceramics, the "first cousin" of glass, are important options for winemakers. Concrete tanks are used rather than stainless steel ones in many wineries, and winemakers wishing to embrace traditional, even ancient, methods are turning to ceramic amphorae similar to those used by their earliest predecessors. Although we focus on storing, shipping, and consuming wine in this book, glass also has various roles around the winery. Chapter 6 provides examples of glassware used in barrel tasting and the wide variety of scientific glassware used in the winery laboratory where chemically durable glass compositions are required (Figure 1.7).

Armed with knowledge of the chemistry of wine, we can more fully appreciate the poetic beverage that finally arrives at our lips after its long voyage encased within glass surfaces. Simple and functional drinking glasses are now available for everyday use, but special occasions call for fine "crystal" with its high sparkle and nearly perfect transparency. The specific technology that provides the "sparkle" (one of the seventeenth century inventions) and the manufacturing steps that produce this elegant glassware will be revealed in Chapter 7 within the context of modern glassmaking. We will explore the inorganic chemistry of glassmaking in contrast to the organic chemistry of winemaking.

One of the seminal developments in glass manufacturing was the development of the blowpipe by the Phoenicians around 200 BC, making glass blowing possible (Figure 1.8). This technique is still used to make some expensive wine glasses and large format Champagne bottles, but most bottles and wine glasses

Figure 1.8 This statue at the Corning Museum of Glass memorializes the art of glass blowing. (Reproduced with kind permission of the Corning Museum of Glass.)

are now manufactured by machine. These mechanical technologies will be illustrated in Chapter 7.

In Chapters 8 through 10, we delve deeper into the inorganic world of glass itself and expand on our discussions of the structure and properties of glass. *Glass* science is a subset of the broader field of *materials* science that has a simple foundational principle: *structure leads to properties*. This concept, borrowed from chemistry and physics, means that we explain the behavior (the "properties" or measurable characteristics) of materials in terms of their small-scale structure. By "small-scale," we mean really, really small from the atomic level through the microscopic. For example, the way in which an empty metal soft drink can is deformed when we squeeze it in our hand can be explained ultimately by the way in which individual atoms of metal move relative to each other.

We can define "atomic scale" in terms of the size of individual atoms: typically a few Ångströms in diameter, with an Ångström equaling one tenth of a nanometer in the metric system of measurement. Much attention has been paid to nanotechnology since President Clinton inaugurated the National Nanotechnology Initiative on the Cal Tech campus in 2000, beginning with the words "Imagine what could be done . . ." The excitement about the science and resulting technology that derives from working at this length scale has not subsided in the time since that launch. This more-than-a-decade time frame is a long one in modern technology, with many hot topics rising and falling in considerably shorter fashion. For those not familiar with the nanometer unit of length, the prefix "nano" means that it is one billionth of a meter, and, for those not familiar with a meter (mostly in the United State, the only industrialized country that has not adopted the metric system as the official system of measurement), a meter is about three feet or one yard in length.

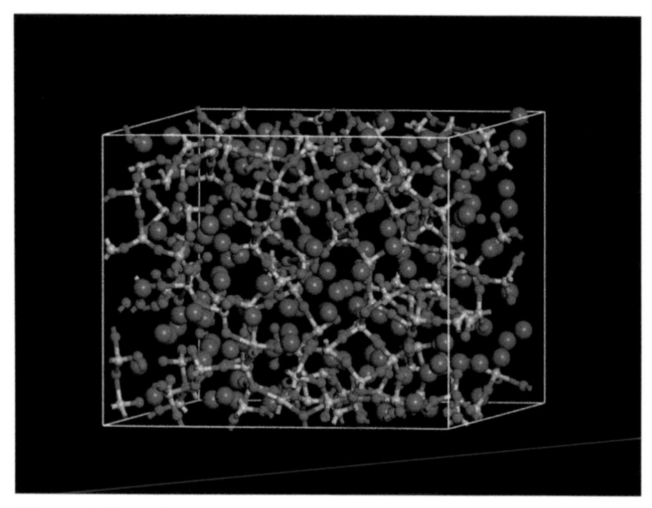

Figure 1.9 This computer-generated image of a sodium silicate glass on the atomic scale shows sodium atoms (large purple spheres) scattered among a rigid network of silicon and oxygen atoms in a "stick-and-ball" configuration (with yellow atoms being silicons and red atoms being oxygens). Such computer models are the focus of Chapter 8. (Reproduced with kind permission of Sabyasachi Sen, Department of Materials Science and Engineering, University of California, Davis.)

So, we will find some properties determined by the nature of the individual atoms that make up the glass and that other properties will be determined by structural features – the size of clusters of atoms (the nanoscale). Chapter 8 will discuss the unique and elegantly disordered atomic-scale structure of glass, with a bonus being the visual beauty of these structures.

Only relatively sophisticated electron microscopes can image individual atoms and nanoscale structures. (The nanoscale is generally defined as a feature being between 1–100 nm in length.) Advances in computing power in recent decades have led to the subfield of computational materials science. The detailed and scientifically rigorous image in Figure 1.9 is produced in this way. When we refer to microscopic-scale structure though, the reference tends to be the conventional optical microscope, with an ability to view structures no smaller than about 1 μm, a small dimension but equal to 1000 nm. The micrometer is often called simply the micron. Such "microscopic" features as micron-scale air bubbles can have significant effects on both mechanical behavior (Chapter 9) and optical behavior (Chapter 10).

In Chapter 11, we discuss the combination of tradition and pragmatics that has emerged to dictate the shape of wine bottles, from the deep "punt" on the bottom of the Champagne bottle to the sloping shoulder of bottles containing pinot noir, the elegant grape of the Burgundy region of France. The deeply indented bottom of the Champagne bottle serves the wholly pragmatic function of containing the internal pressure while giving waiters worldwide the ability to pour the sparkling wine into their guest's glasses with one-handed panache. The

Figure 1.10 These elegant wine glasses from the eighteenth century were a symbol of sophistication and fine dining. (Reproduced with kind permission of the Corning Museum of Glass.)

Burgundy's sloping shoulder is a signature shape for pinot noirs worldwide. On the other hand, the square shoulder of bottles for the red wine of Bordeaux is pragmatic, facilitating the removal of sediment as old and elegant cabernet sauvignons are slowly poured. A discussion of wine bottles is not complete without the related technology of labels, both pragmatic sources of consumer information and, increasingly, canvases for artistic expression. In some cases, glass decoration is replacing traditional paper labels.

The technology of drinking glass manufacturing continued to evolve through the eighteenth and nineteenth centuries when elegant designs produced with meticulous craftsmanship became an essential part of the dining experience in upper class homes in both Europe and the New World (Figure 1.10). We will see in Chapter 12 that, during that time, the focus was generally on the ornate as an unambiguous symbol of the elegance of wine appreciation.

The twentieth century brought a revolution in design with wine glasses being a prime example. The ornate gave way to elegant simplicity. In addition in the last half of the twentieth century, the shapes of wine glasses developed unique relationships to the wines contained within. For example, the Champagne flute avoids the rapid dissipation of the effervescent bubbles and delivers the wine directly to the back of the tongue. The larger glasses for Burgundies (pinot noirs) provide more surface area exposed to air, releasing more intense aromas especially when the wine is swirled. We find in Chapter 12 that, since the 1950s, more than 2 dozen varietal-specific stemware shapes have been developed. Wine glass manufacturers and many connoisseurs have made a convincing case for serving pinot noir in stemware with a distinctly different shape from that optimized for cabernet sauvignon that, in turn, is distinct from that for sauvignon

Figure 1.11 Contemporary wine glasses come in a variety of sizes and shapes deemed appropriate for a wide array of varietals. (Reproduced with kind permission of Kim Goodwin, Luigi Bormioli Italy.)

blanc, and so on (Figure 1.11). The claim is simply that, when tasting wine, the shape of the glass affects the bouquet, taste, balance, and finish. Research primarily done by the Riedel family of the famous Austrian wineglass company involved evaluation of a wide range of varietals in a wide range of shapes by professional tasters to settle on the optimal matches. Recently, a number of important wine writers have begun to rebel and recommend instead a one-size-fits-all shape to simplify life for connoisseurs (and themselves).

Chapter 13 reminds us that cork has been the dominant technology for sealing wine bottles for roughly five centuries and has made the international appreciation of wine possible, but alternate technologies now challenge that dominance. A significant lapse in quality control in the Portuguese cork industry in the 1980s accelerated the development of these alternatives. Synthetic corks of various types have been used, and metallic screw caps have become especially popular in the New World markets of Australia and New Zealand. Another of the challengers to cork is the glass stopper that has an interesting history from its traditional role in the scientific laboratory to substantial research and development in the past two decades, specifically for the purpose of sealing wine bottles. These stoppers are widely used for German white wines and are finding increasingly wide acceptance around the world. Could the wine bottle of the future be all glass all the time? Well, not exactly. The success of the glass

Figure 1.12 Decanters are sometimes helpful for aerating wine and to minimize sediment. This is especially true for powerful tannic wines such as the Barolos from the Piedmont region of Northern Italy.

stopper rests on a small, polymeric o-ring. Nonetheless, the *nearly* all glass bottle stopper system is now an important option for the wine industry.

Despite the romance surrounding the long-term cellaring of "collectable" wines, most wine is consumed shortly after purchase. Whether or not the consumer stores the wine for many years "in bottle," the final enjoyment of this nectar of the gods and goddesses is invariably in glass stemware, and, just prior to drinking, the wine is sometimes transferred from the bottle to glass carafes or decanters (Figure 1.12). This is especially true for older, tannic wines. Just as the Bordeaux bottle shape facilitated the removal of sediment in cabernet sauvignon, pouring the contents of the bottle into a decanter can also allow the settling of the sediment before pouring. Equally important, pouring wine out of the bottle and swirling it in the decanter can provide substantial aeration. The relationship of wine to oxygen is fascinating and complex. Isolating wine from oxygen is a major role of cork (and metal screw cap and glass stopper) bottle closures. On the other hand, when the bottle is subsequently opened, time for exposure to air may be required to optimize the wine's character. Those sediment-laden tannic wines often benefit from this oxygenation turning a harsh wine straight from the bottle into a balanced and refined masterpiece. There is considerable art and science in the preparation of wine in many different decanter designs, as we shall see in Chapter 14.

Finally in Chapter 15, we critically explore whether the ubiquitous role of glass in wine appreciation is secure or if glass bottles and stemware will go the way of the chrome bumpers on automobiles of a half-century ago. Of course, it is hard to imagine ordering a fine wine in a Michelin-starred restaurant and

Figure 1.13 Glass stemware enhances the experience of fine dining. All of these varied glasses can find their way to the table when experiencing the full tasting menu at the Quince Restaurant in San Francisco, California. (Reproduced with kind permission of Zoe Simonneaux, Quince Restaurant.)

having that wine arrive at the table in anything other than a glass bottle. Similarly, we expect the table to be set with elegant glass stemware (Figure 1.13). So while the chrome bumper analogy may be far fetched, the adjective "ubiquitous" could be the endangered species, as *vin ordinaire* is increasingly available in packaging imitative of the beer, soft drink, and milk industries.

Throughout these 15 chapters, our goal will be to enhance the readers' appreciation of wine by better understanding the vessels from which that wine is delivered to their lips. So, now let us begin by exploring the history of wine.

BIBLIOGRAPHY

Callister, William and David Rethwisch, *Materials Science and Engineering: An Introduction*, 9th Edition, John Wiley & Sons, Inc., New York (2014).

Clinanti, Pino, *El Fuoco, Il Vetro, Il Vino*, Fondazione Banfi, Montalcino (Italy) (1992).

Ellis, William S., *Glass*, Bard/Avon, New York (1998).

MacNeil, Karen, *The Wine Bible*, 2nd Edition, Workman, New York (2015).

McGovern, Patrick, *Uncorking the Past*, University of California Press, Berkeley (2009).

Shackelford, James, *Introduction to Materials Science for Engineers*, 8th Edition, Pearson, Upper Saddle River, NJ (2015).

Taber, George M., *Judgment of Paris: California vs. France and the Historic 1976 Paris Tasting that Revolutionized Wine*, Scribner, New York (2005).

Taber, George M., *To Cork or Not to Cork: Tradition, Science, and the Battle for the Wine Bottle*, Scribner, New York (2007).

Chapter 2

A Brief History of Wine – Storing and Drinking Wine Before Glass

"One of the disadvantages of wine is that it makes a man mistake words for thoughts."

Samuel Johnson as quoted in *The Life of Samuel Johnson, LL.D.*, Volume 2 by James Boswell (1791)

Louis Grivetti had it right. The Emeritus Professor of Food Science at the University of California, Davis has written about wine as "the food with two faces." Our first two chapter opening quotations reveal the two faces that emerge from the fermented grape. In Chapter 1, we celebrate the best from wine: courage, love, and joy! But opening this chapter, the sage Dr. Johnson reveals the other face, the drunken one (Figure 2.1). How many relationships have been affected – sometimes tragically so? The history of wine is the story of a journey that reveals both faces from time to time. While we focus on a celebration of our favorite beverage, every stop along the historical trail is not filled with inspiration and joy, but each is interesting. So, let us begin.

Patrick McGovern of the University of Pennsylvania has done as much as anyone to pursue the early history of producing fermented beverages. His many books, including *Uncorking the Past: The Quest for Wine, Beer, and Other Alcoholic Beverages* (University of California Press, Berkeley, 2009), have chronicled our continuing affection for fermented beverages in general and wine in particular. He has had a significant impact on the collections and exhibitions at the University of Pennsylvania Museum of Archeology and Anthropology (the Penn Museum), about which he says "The whole museum is very much a

The Glass of Wine: The Science, Technology, and Art of Glassware for Transporting and Enjoying Wine, First Edition. James F. Shackelford and Penelope L. Shackelford.

Figure 2.1 Caravaggio's famous portrait of Bacchus, c. 1589, illustrates a half-drunk teenager with a remarkable skill for steadying a glass of wine. Bacchus has become an iconic image of the excesses of wine, but the beginnings of wine appreciation (and excess) preceded the availability of glass vessels. (Reproduced with kind permission from Alamy.)

paean to mankind's love affair with fermented beverages." An impressive example is given in Figure 2.2.

McGovern's substantial knowledge about ancient wine and other fermented beverages comes from what he terms "biomolecular archeological investigations." He and his colleagues use a wide-ranging suite of analytical tools for characterizing the chemistry of these ancient beverages and their containers. From liquid chromatography–mass spectrometry (LS-MS) to carbon and nitrogen isotope analysis, all their techniques provide chemical "fingerprints" of the components. Modern DNA testing is similarly useful, especially in regard to identifying grape and cereal sources for the wine in comparison to modern counterparts as well as the historical migrations of the ancient wine "connoisseurs." Of course, old-fashioned visual inspection and optical microscopy are also available for their archeological detective work. We need not be masters of these technologies ourselves to appreciate that McGovern and his colleagues have

Figure 2.2 This ceramic wine amphora at the Penn Museum was retrieved from a Mediterranean shipwreck in the early 1950s by Jacque Cousteau. (Reproduced with kind permission of Penn Museum.)

used a powerful range of scientific instruments to provide us with their insight as to the nature and the evolution of wine culture.

Vernon Singleton, a Professor in the Department of Viticulture and Enology at UC Davis, coauthored with his colleague Maynard Amerine the seminal book *Wine: An Introduction for Americans* in 1967, helping to elevate wine consciousness as the California wine industry was in ascendance. He has also provided the wine chemist's perspective on the key pieces of evidence that can be gleaned from the suite of instruments used by McGovern, et al. The great challenge of course is to find a wine containing vessel intact after millennia, let alone the wine itself. Wood and animal skin materials generally degrade, but pottery, stone, and glass often survive. These more durable vessels, however, are seldom able to retain their contents indefinitely especially in light of sealing technologies available to the ancients. The best case we can expect is desiccation in an arid location. In this case, Singleton suggests that one of the best candidates to indicate the residual presence of wine are tartaric acid and its salts that are most likely to come from grapes rather than any other common fruit. Syringic acid is also a good indicator that is derived from the alkaline fusion of residue from red wine's anthocyanin pigments and tannins. The scientific instruments already listed are especially effective for identifying these analytes.

Science can do more than looking backward in time, gleaning the presence or absence of wine in ancient vessels. Some other instruments are helping us understand the effect of alcohol on the human brain. While inserting probes into the brain is not permitted for humans, electroencephalographs can monitor brain waves while radioactive markers can be injected into volunteers to trace the path of alcohol through the blood – brain barrier. Even more dramatic, noninvasive imaging techniques show great promise for detailing which regions of the brain are activated by a given beverage. Functional magnetic resonance imaging (f-MRI), positron emission tomography (PET), and single-photon emission computed tomography (SPECT) are all being used for this purpose. Current research promises to give fundamental understanding of why we sometimes respond to wine with the uplifting words opening Chapter 1 and sometimes behave more like the tipsy Bacchus of Figure 2.1.

Penn's Dr. McGovern received considerable, well-deserved publicity in 2004 for the discovery of the "earliest alcoholic beverage," a mixed fermented drink of rice, honey, and hawthorn fruit and/or grape found on Chinese pottery shards dated 7000–6600 BC. These pottery jars were from the Neolithic village of Jiahu, in the Henan Province of North-Central China. Evidence nearly as old has placed barley beer making and grape winemaking in the Middle East. Pottery dating from 6000 BC allowed Mediterranean people to store wine. McGovern notes that the invention of pottery in this 6000 BC time frame facilitated the "settling down" of people and allowed wine (and food) to be prepared and stored in vessels made of ceramics and animal skins. He also says we should not be surprised if some new discovery in the Middle East edges out the Jiahu sample for oldest fermented beverage. Certain villages in eastern Turkey are prime candidates. Specifically, McGovern and Swiss scientist José Vouillamoz have identified evidence of winemaking in southeastern Anatolia, in the Asian portion of modern Turkey, sometime between 5000 and 8500 BC raising the possibility of wine production there prior to 7000 BC In any case, a view of the Cappadocia region of Anatolia known to contain carved out spaces for ancient wineries is shown in Figure 2.3. The "Neolithic revolution" occurred across many cultures in this time frame bringing the advent of farming and culminating with the widespread use of metal tools in the **Bronze** and **Iron Ages**. (**Bronze** is a metal **alloy**, that is, a metal composed of more than one element; in this case, an alloy of primarily copper and tin.)

In the Chinese study, McGovern and his colleagues also reported the analysis of remarkably preserved liquid samples in tightly sealed bronze vessels that proved to be rice and millet wines, flavored with herbs, flowers,

Figure 2.3 (a) The austere, even surreal beauty of the Cappadocia region of Anatolia (located in modern Turkey) was created by the erosion of soft tuff stone produced by the consolidation of volcanic ash. (b) The carved out openings created by ancient settlers lead, in some cases, to rooms that served as wineries. Although primitive, wine storage would have been convenient in such "natural" cellars.

and/or tree resins similar to oracle descriptions of herbal wines in the Shang dynasty (Figure 2.4). The press frenzy around the McGovern disclosure was amplified by the fact that the conventional assumptions about the origin of wine centered on the Middle East. As just noted, the Chinese discovery correlated rather well with the evidence for winemaking in the Middle East from the same early Neolithic period (7000–5000 BC).

What is most intriguing about this competition between these sites on either side of Asia is their substantial separation. McGovern notes that some communication might have been possible along a prehistoric predecessor of the Silk Road, but he actually leans toward another hypothesis, that of the "drunken monkey." In essence, humans and their ancestors have been, by and large, sweet fruit and berry eaters and the natural fermentation of these materials has developed a genetic predisposition in us for such beverages. Our ancestors would have been expected to consume moderate amounts of alcohol as a by-product of consuming such fruits and berries with the attendant health benefits that have become the focus of much research in recent times. In other words, the genesis of wine was likely not within a particular cultural context but in our DNA! Scientists have specifically focused on the ADH4 gene, finding that it mutated about 10 million years ago in a common ancestor of humans, chimpanzees, and gorillas creating a more efficient version of the enzyme that metabolizes ethanol. Those of *us* with this version would then have a wider range of resources to gather, including fruits that might occasionally have a delightful level of fermentation.

Harold Olmo, the late and great viticulturist at UC Davis, noted that a special challenge for studying ancient wines is the rapid rate of disappearance of

Figure 2.4 Dr. Patrick McGovern of the University of Pennsylvania Museum of Archeology and Anthropology (the Penn Museum) samples the "nose" of a 3000-year-old millet wine that had been preserved inside a tightly sealed bronze vessel from the Shang Dynasty in China. (Reproduced with kind permission of Thomas Stanley, University of Pennsylvania Museum of Archaeology and Anthropology.)

wild vines. While the *Vitis vinifera* (grape) is one of the oldest cultivated plants that provide living progenitors, its origin and the nature of its domestication is a challenge. Nonetheless, Olmo provided a relatively full picture of how domestication could have occurred. He expected that people selected particularly fruitful vines in native forest habitats. These wild species are often classified as *Vitis sylvestris* and are dioecious with male and female flowers on separate plants requiring pollination. (*Sylvestris* refers to the forest.) Moving the wild vines to adjacent, cleared areas propagated the vines. It is interesting to note that the sporadic occurrence of self-propagating (hermaphroditic) traits in the wild populations of the southern Caspian Sea area of the Mediterranean basin became the norm in cultivated varieties. These derived hermaphrodites were rapidly disseminated. So, we can thank self-fertilization for the migration of the vines across the ancient world (Figure 2.5).

Figure 2.5 This contemporary descendent of the *Vitis vinifera* (grape) can thank hermaphroditic behavior for its abundance in today's world.

This migration allowed Chinese poets to celebrate the mystical pleasures of fermented beverages in eastern Asia while Arabic and Persian Bacchic poets did so in the west, although with an additional, erotic theme. These poetic celebrations often referred to mixed beverages, such as the Jiahu discovery, but, by and large, grape wine dominated in western Asia and cereal wine in the East.

Given today's international cultural wars, there is considerable irony that one of the earliest pieces of chemical evidence for unadulterated grape wine is the Hajji Firuz beverage (dated 5400–5000 BC) from ancient Persia, the land that is now Iran, on the western periphery of the prehistoric Silk Road (Figure 2.6). The near absence of wine in current Iranian culture does not erase a rather rich history of wine consumption in the region. High-quality wines were also produced in the ancient land of Canaan (including Phoenicia in its northernmost part) in the same period from the sixth millennium BC onward. Patrick McGovern estimates that this "fine wine" making was done in the large fertile valley between the Lebanon and Anti-Lebanon mountains. Herodotus, the Greek who lived in the fifth century BC, is considered to be the "Father of History," the first to systematically record his observations and arrange them in a compelling

Figure 2.6 This vessel contains some of the earliest evidence of wine made strictly from grapes. The wine jar is from the Hajji Firuz Tepe, an archeological site in northwestern Iran. (Reproduced with kind permission of Penn Museum.)

narrative. His account of the Greco-Persian Wars that took place a few millennia after the pioneering Hajji Firuz wine dwelled on the fact that the Persians were fond of deliberating on weighty affairs while drunk, but they wisely reviewed their decisions the following day when sober.

While grape-based winemaking evolved in Persia and the rest of the Middle East, the period of 3000–1500 BC saw Egypt as the world's superpower and, as a by-product of that eminence, the center of wine culture. Just prior to this reign of enological eminence, the tomb of King Scorpion I (around 3150 BC) included an early version of a wine cellar with 700 wine jars. As with many ancient wines, these beverages were preserved with resins from pines and other trees. They were also sweetened with whole raisins. This early Egyptian wine, however, was likely imported from the Canaan (especially the Phoenician) area. Egypt was not an ideal venue for grape growing. Nonetheless, within two centuries (in Dynasty I as compared to Dynasty 0 of Scorpion I), the pharaohs of a united Egypt established a royal wine industry in the Nile Delta. The Canaanites were brought in as consultants to set up these vineyards and wineries. Many of these vintners had Semitic names, and distinctive images remain in Egyptian tombs. Figure 2.7 shows a nineteenth century illustration of Egyptian winemaking technology.

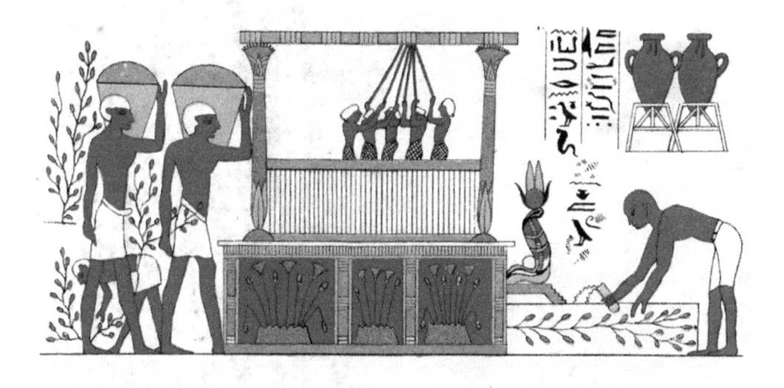

Figure 2.7 This 1864 lithograph illustrates the winemaking process as established by the Egyptian pharaohs in the Nile Delta from about 3000 BC. (Reproduced with kind permission from Alamy.)

Although the climate of Egypt was not optimal for winemaking, the alluvium of the Nile Delta was a calcareous soil surprisingly similar to parts of the modern Bordeaux region. By 2000 BC, after a thousand years of winemaking, wine had become a significant part of Egyptian society but still could not match the popularity of beer. The goddess Hathor (Figure 2.8), associated with a fondness for red beer, was sometimes known as "the mistress of drunkenness." Aside from cultural tastes, there was a distinct advantage to beer making in that cereal could be stored until there was a convenient time for fermentation. The same was not true for grape storage in the warm to hot climate of Egypt. Nonetheless, the Egyptians carried on the winemaking skills of their Canaanite mentors and are credited with discovering that aging can improve wine, as they observed the effect of wine storage in some of their amphorae.

Figure 2.8 This statue of the head of the goddess Hathor is in the Dandera Temple complex in Egypt. With her distinctive cow's ears, Hathor was a symbol of both fertility and a fondness for alcohol. (Reproduced with kind permission from Alamy.)

The Canaanites also ventured on to Crete, where they laid the foundation for Greek winemaking and then to continental Europe. The Greeks commonly had "mixed" fermented beverages such as Kykeon (Greek for "mixture"), a combination of wine, beer, and mead (fermented honey). The beginning of the Iron Age (around 900 BC) saw large-scale wine production in Italy. The coastal cities in what is now modern Calabria, in the toe of Italy, was once known as Oenotria, the land of trained vines. Today at Pompeii, at the base of Mt. Vesuvius, the Mastroberardino Winery is making wine from ancient Roman varieties, such as Greco di Tufo, in traditional ways, including the use of Roman-style trellises.

As the Canaanites, in general, had tutored the Egyptians, those from Phoenicia likely tutored the Etruscans. By 600 BC, the Etruscans became the principal exporters of wine to southern France, and a wine culture migrated further north to Trans-Alpine Burgundy and the Mosel. Beyond that point, however, Nordic grog prevailed. The Phoenicians' largest colony to the west was Carthage on the North African coast in what is now Tunisia. Wine was the beverage of choice there. Their "raisined wine" is, in fact, similar to the modern ripasso (meaning "repassed") wines of northern Italy where the pomace of leftover grape skins and seeds from the fermentation of amarone grapes are added to the batch of Valpolicella wines for a period of extended maceration boosting the alcohol level and body of the wines while leeching additional tannins. The resulting complexity, flavor, and color are substantially enhanced. Carthage's influence spread across the Mediterranean to Spain's Costa del Sol with a heritage worthy of the powerful wines now coming from the Priorat region of southeastern Spain. These contemporary heirs to Phoenician winemaking technology are shown in Figure 2.9.

Figure 2.9 The ripasso wine from the Valpolicella region near Verona, Italy and the Priorat wine from southeastern Spain are contemporary heirs to the Phoenician winemaking techniques introduced in North Africa more than two millennia ago.

We have now seen the genesis of winemaking in the three great wine producing nations of modern Europe: France, Italy, and Spain. We should also appreciate the history of fermented beverages in the New World. The current, substantial wine industry in North and South America is a story of migration from these European traditions. Long before the European influence, however, the Mapuche in Chile chewed fruits from shrubs and trees producing fermented *chicha*. The domestication of maize systematically led to the production of corn beer, also known as *chicha*, the generic Spanish term for any fermented beverage from the Americas. *Chicha* played an important role in the social and religious life of the later Incan empire of Peru. And, the domestication of the cacao tree gave us both chocolate and a mildly alcoholic beverage (estimated at 5–7% alcohol) made by Pacific coast Guatemalans from fermented cacao pulp in dugout canoes.

Humans seem to be ready and willing to make a fermented beverage from any starting material from grapes to barley to cacao, but notice that all of the examples just given were in *South* America. Archeological evidence indicates that North American natives were teetotalers. McGovern suggests that this New World picture is further evidence of sorts for the drunken monkey hypothesis. The North American ancestors were from Siberia where there was a shortage of high-sugar resources. Thus, they did not have the predisposition to produce fermented beverages. On the other hand, the use of hallucinogenic mushrooms by Siberians might suggest a broader "stoned monkey" hypothesis for humankind's desire for any sort of chemical escape from the challenges of everyday life.

This worldwide scan of the history of winemaking has lacked the need to mention the role of glass. We can now turn to the considerably younger history of this material to see when it finally entered the field of enology in a significant way, edging into the timescale of this chapter and then steadily growing in importance until today's role as the dominant material for storing, shipping, and consuming wine.

BIBLIOGRAPHY

Amerine, Maynard and Vernon Singleton, *Wine: An Introduction*, 2nd Edition, University of California Press, Berkeley (1977).

Grivetti, Louis E., "Wine: The Food with Two Faces," in *The Origins and Ancient History of Wine*, P.E. McGovern, S.J. Fleming, and S.H. Katz, Eds., Gordon and Breach, New York (1996), pp 9–22.

MacNeil, Karen, *The Wine Bible*, 2nd Edition, Workman, New York (2015).

McGovern, Patrick, *Uncorking the Past*, University of California Press, Berkeley (2009).

Olmo, Harold P., "The Origin and Domestication of the *Vinifera* Grape," in *The Origins and Ancient History of Wine*, P.E. McGovern, S.J. Fleming, and S.H. Katz, Eds., Gordon and Breach, New York (1996), pp 31–43.

Singleton, Vernon L., "An Enologist's Commentary on Ancient Wines," in *The Origins and Ancient History of Wine*, P.E. McGovern, S.J. Fleming, and S.H. Katz, Eds., Gordon and Breach, New York (1996), pp 67–77.

Chapter 3

A Brief History of Glass – and How It Came to Dominate Wine Appreciation

"What moved me most, however, . . . illuminated glass objects which had been pulled up from the depths – cups, bowls, pitchers, perfume flasks, goblets . . . when the doors close and it grows dark, the curator presses a switch turning on small light bulbs inside the little vessels, bringing to life the fragile, matte pieces of glass, which start to sparkle, brighten, pulsate. . . . We stand in darkness surrounded by light."

Ryszard Kapuściński on his experience in the Museum of Underwater Archeology in Halicarnassus where the Greek historian Herodotus was born, in *Travels with Herodotus* (Vintage Books, New York, 2007)

And, so Ryszard Kapuściński, modern-day historian and journalist, was entranced by the mystery and beauty of *glass*, by glass vessels whose purpose in ancient times was, among others, the storage and consumption of wine. Divers had retrieved these enchanting fragments from the bottom of the Aegean Sea. Our previous chapter identified the origins of wine as a few millennia before glass was invented and several millennia before Herodotus. While glass is a more recent human discovery, we cannot call it an entirely human invention, as nature has produced glass objects as long as lightning strikes have occurred in the vicinity of sand deposits, both in deserts and seaside beaches. Nonetheless, evidence has been found for human-made glass in ancient Mesopotamia (comparable to modern Iraq and northern Syria) dating from before 2000 BC. By that time, wine had been made for at least three millennia throughout Asia, and the leading winemakers were now in the superpower of the day, Egypt.

The Glass of Wine: The Science, Technology, and Art of Glassware for Transporting and Enjoying Wine, First Edition. James F. Shackelford and Penelope L. Shackelford.

Figure 3.1 Ancient clay tablets like this one provided formulas used by the earliest glassmakers in Mesopotamia. (Reproduced with kind permission of the Corning Museum of Glass.)

An ancient cuneiform text from Mesopotamia gave a recipe: "Take sixty parts sand, a hundred and eighty parts ashes from sea plants, five parts chalk, heat them all together, and you will get glass." Figure 3.1 shows a ceramic tablet used for such recipes. The Mesopotamian evidence has supplanted a popular legend from Pliny the Elder that placed the discovery of glass in the first century AD. In this story from his *Natural History*, Phoenician sailors had come ashore to prepare an evening meal on the beach. Unable to find any stones on which to rest their pots, they used some lumps of material from the ship's cargo, a natural form of soda called natron, a material at the time used in embalming the dead. As the fire heated up, the natron began to react with the sand, and molten glass flowed before the sailor's eyes like lava emerging from a volcano. Pliny's story may well be true, but that molten glass was not humankind's first glassy achievement.

Just as the honor of "oldest fermented beverage" or "first winemaker" is continually subject to the findings of the latest archeological discovery, the location of the "first glassmaker" might drift around even more in the future. Nonetheless, all evidence and legends to date have shown that the way glass is made has changed little from the earliest days up to now as it has evolved into a multibillion dollar industry and a core part of modern infrastructure, providing enough energy to convert locally available deposits of sand (SiO_2), soda ash (Na_2CO_3), and lime (CaO) into a durable and transparent material. Over time, the combination of these three compounds has come to involve mostly sand (generally between two-thirds and three-quarters) with the remainder roughly split between the soda ash and lime. Melting them together in a furnace breaks down the soda ash into soda (Na_2O) and gaseous carbon dioxide (CO_2). The bubbles of CO_2 gas play an important role, mixing together the remaining ingredients into a homogeneous blend known as **soda-lime-silica glass**, where the **silica** is the SiO_2 that comes from common sand deposits.

There are important environmental issues tucked away in this glass chemistry. The massive quantities of glass produced for modern society mean that the CO_2 produced when soda ash decomposes is released in massive quantities. The glass industry is thus a significant contributor to concerns about global climate change related to human-produced CO_2. (An even bigger concern is the cement industry in which even greater quantities of limestone, $CaCO_3$, are decomposed to produce quicklime, CaO, producing much greater volumes of CO_2.) Beyond CO_2 release, the glass industry is a major energy consumer. A typical soda-lime-silica glass is melted at temperatures above about 1200 °C and formed into final product shapes at temperatures above 700 °C. These high processing temperatures for such large quantities of products make the glass industry one of the leading consumers of energy in an energy-hungry industrial society.

Glass technologists use the term "fining" to describe the process by which CO_2 gas bubbles homogenize the glass melt. The availability of increasingly high-temperature furnaces in the twentieth century made this fining process increasingly effective so that we now routinely expect windows, wine glasses, and other glass products to be clear and bubble-free. Older glass products such as nineteenth-century windows we might find in an old home have a characteristic "bubbly" appearance. Those windows were made in lower temperature furnaces, and the fining bubbles could not escape from the relatively viscous glass melt. Higher temperature furnaces decrease the melt viscosity substantially and permit the fining bubbles to escape, not unlike Champagne bubbles escaping from a warm glass of Champagne.

Before we go further with analogies of Champagne and the inevitable image that conjures up an elegant and perfectly transparent glass flute, we should return to the reality of glassmaking at the beginning, in Mesopotamia and as the craft was transmitted on to Egypt, not unlike the journey of winemaking some millennia earlier. In those first years of glassmaking, the common method was "core forming" in which glass was molded onto an inner core generally made of clay. It is not entirely clear how the glass was applied to the core: by laying strips of

Figure 3.2 These colorful vessels from Egypt were produced in the time frame of 1450 to 1150 BC. (Reproduced with kind permission of the Corning Museum of Glass.)

molten glass onto it, dipping the core into a batch of molten glass, or coating the core with powdered glass and then heating the powder/core system together to form the glass vessel. In any case, brightly colored core-formed drinking vessels and perfume bottles have been identified in Egypt and dated over the 300-year time period between 1450 and 1150 BC (Figure 3.2).

The wine-like migration of glass technology continued as glass became a prized possession in the Roman Empire. Glass beads were relatively easy to fashion and served as both substitute and peer for precious stones. With time, the Roman glass vessels were more broadly available and served more than the wealthy for purposes as diverse as drinking, food storage, and burying ashes of the dead. We now have evidence of nearly two-dozen wine bars in Pompeii along a single road, with glass as the vessel of choice. Patrons often drank from still unrefined vessel shapes with rounded bottoms unable to stand upright on their own.

Legend has it that even the Holy Grail of Christ's Last Supper might have been a product of Roman glass craftsmanship. Some believers claim that a dark green glass vessel in the Museum of the Cathedral of San Lorenzo in Genoa, Italy, is, in fact, that sacred object and is known there as the *Sacro Catino*, brought back by Genoese sailors from Palestine. Chemical analysis of a small fragment of the vessel in the 1950s at the Brookhaven National Laboratory in New York, however, indicated that impurity levels of magnesium oxide and potassium oxide were much too large to be consistent with traditional Roman glass recipes. This sort of chemical signature is as important in identifying ancient glass samples as Patrick McGovern's biomolecular analysis was in identifying ancient wines in the previous chapter. Nonetheless, this scientific doubt has not diminished the faith of many of the Genoese believers.

The wider availability of glass with time in the Roman Empire can be attributed to the most important development in glass technology after the discovery of glassmaking itself, namely, the invention of the blowpipe by Phoenicians around 200 BC. The ability to form bulbous shapes of molten glass on the end of a long tube by blowing into it and then further forming the glass piece with various

Figure 3.3 A modern glassblower demonstrates the ancient craft on the Hot Glass Show stage at the Corning Museum of Glass in Corning, New York. (Reproduced with kind permission of the Corning Museum of Glass.)

shaping tools provided an economical manufacturing technique (Figure 3.3). The blowpipe remained a standard technology for glassmaking until the industrial revolution of the nineteenth century when machines became available for the same purpose and with even greater economy.

With the coming of the Renaissance, the unrivaled center of glassmaking excellence moved north from Rome to Venice, especially within the secretive guilds on the island of Murano. Glass factories were relocated from Venice to Murano in the late thirteenth century as the high density of furnaces constituted a serious fire hazard for the city itself. The legendary secrecy is explained by the fact that industrial espionage was a concern even in those days. A *capitolare* or book of ordinances provided government oversight of the highly regarded glass artisans there. The compact space of the island also aided in maintaining the secrecy of the craft. At times, the insistence on secrecy was a bit heavy-handed. Assassins were dispatched to carry out death penalties on rogue artisans who dared to practice their trade outside Murano. As oppressive as it might have been, the intense environment on Murano bore results. The zenith of Murano's output came with its elegant *cristallo*, a thin and "crystal clear" glass in which the "lime" of soda-lime glass was replaced with a potassium oxide-rich vegetable ash. Thin and fragile material was especially responsive to the creative work of the skilled glassblowers yielding goblets that were among the triumphs of the Murano craftsmen. Many of these vessels for wine and other beverages were highly decorated with enamels and opaque glass fibers; some were colored. Ice glass had a shimmering and shattered appearance that resulted from immersing the hot glass into cold water.

The first known textbook on glassblowing came from another part of Northern Italy, Florence. Antonio Flori published *L'Arte Vetraria* in 1612. Glassmaking skills could not be confined to Murano, and, assassins notwithstanding,

Figure 3.4 In this example of George Ravencroft's early lead crystal, a rare goblet has mold-blown ribbing that is pinched to form a mesh design on the bowl. Also, one of the small pieces of glass (prunts) molded onto the stem is stamped with a raven's head seal taken from his family's coat of arms. The prunts provide decoration as well as a firm grip for the goblet. (Reproduced with kind permission of the Corning Museum of Glass.)

several Murano artisans migrated to various parts of Europe. The next great advance in glassmaking also occurred in the seventeenth century not in Italy but in England, from a successful import/export businessman with substantial business in glass who had lived for a while in Venice. George Ravenscroft discovered that adding a substantial amount of lead oxide to glass produced many benefits and a product substantially easier to form and considerably more durable than Murano's *cristallo*.

Aesthetically too, Ravenscroft's glass was superior (Figure 3.4). The lead oxide eliminated the tendency of clouding from the fuming of the sodium carbonate in the glass mixture. We can recall the comment in Chapter 1 that the emergence of highly transparent glass appears to have inspired the winemakers of Champagne to make increasingly high-quality, clear wine that we now routinely associate with the experience of enjoying sparkling wines.

Equally important, lead oxide provides a very high *index of refraction*, or the tendency of light to bend when passing from air into the glass. This term will be described in more detail in Chapter 10 that deals with optical behavior. This refraction is fundamental to the design of focusing lenses such as eyeglasses. The high-atomic number lead has a corresponding high index of refraction. This strong tendency to bend the light path within the glass gives it the highly prized sparkle of this fine "lead crystal."

Of course, the prestige of lead crystal has persisted to this day, but the descriptor *lead crystal* is an unfortunate one. *Lead* is accurate and appropriate, but *crystal* is misleading. As noted above, the great Murano glass was called *cristallo* because it was crystal clear. Similarly, before lead crystal came along, some called all high-quality clear glass "crystal" for its similarity to pure crystalline **quartz** that was sometimes known as "mountain crystal." But describing fine lead glass as crystal is largely based on the similarity in appearance (that sparkle due to the high index of refraction) between this glass and crystals, especially fine gemstones such as diamond. The fundamental nature of glass, however, is that it is *not* crystalline. Glasses are sometimes referred to as **noncrystalline** solids.

The atoms in a **crystalline** solid are arranged in a regular and repeating pattern. Those in a noncrystalline solid are more random, although local chemical bonding produces a regular "building block" geometry. For example, quartz is the most common crystalline form of silicon dioxide and the basis of the structure of ordinary beach sand on the atomic scale. And, as noted above, sand deposits are the source of the major ingredient in most glasses. At the local atomic level, four oxygen atoms surround each silicon atom in a three-dimensional tetrahedral pattern. This structure can be represented in two-dimensional space by a triangular pattern (Figure 3.5).

When we move on to the three-dimensional structure, we find the tetrahedral building blocks in quartz are arranged in a pattern surprisingly similar to the double helix geometry of DNA (Figure 3.6). In fact, the double helix of quartz was known a few decades before the famous DNA discovery of Watson and Crick. It is intriguing to reflect on the fact that nature has chosen the double helix structure for both the building block of life and the most common inorganic material in the earth's crust.

When quartz melts and then is cooled fast enough to prevent it from recrystallizing, the result is glass. The three-dimensional linkage of tetrahedra is random

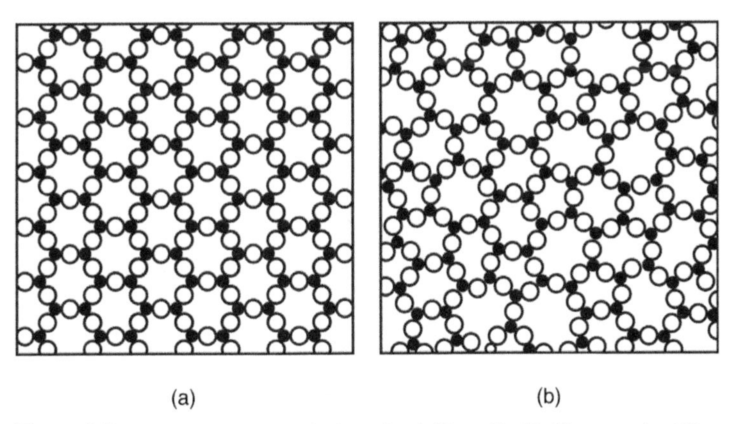

Figure 3.5 This two-dimensional schematic of silicon dioxide illustrates the difference between (a) a crystal and (b) a glass. The atoms in the crystal are arranged in a regular and repeating pattern while those in the glass appear much more random. In fact, the "local" structure is the same, namely, three open circles (representing oxygen) surround each closed circle (representing silicon) in a triangular pattern. (Shackelford, 2015. Reproduced with permission of Pearson.)

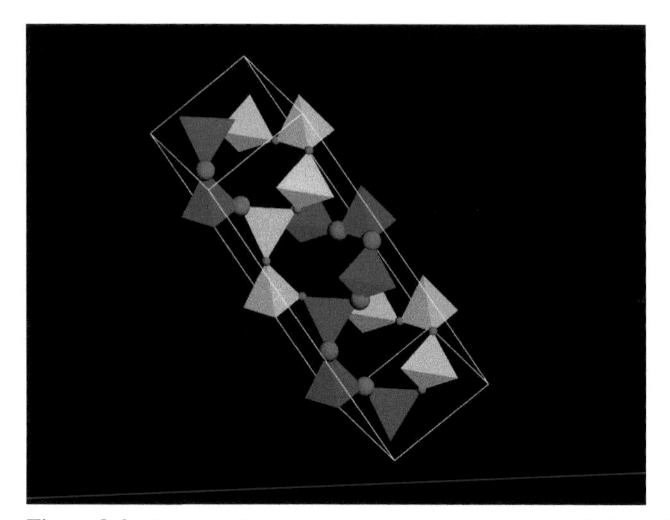

Figure 3.6 The three-dimensional structure of quartz shares the double helix geometry of DNA. Because four oxygen atoms surround each silicon atom, each such "building block" is represented by a tetrahedron. One helix is shown as purple tetrahedra connected by oxygen ions shown as large red spheres. The other helix is shown as yellow tetrahedra connected by oxygen ions shown as small red spheres. (Reproduced with kind permission of Sabyasachi Sen, Department of Materials Science and Engineering, University of California, Davis.)

(Figure 3.7). And, so George Ravenscroft's crystal is not the crystallographer's crystal, but who is to say which atomic-scale structure, the perfect crystal or the perfectly random glass, is more beautiful? We will explore this random beauty in greater detail in Chapter 8.

Nomenclature aside, Ravenscroft's breakthrough owed considerable debt to the British Navy. Admiral Sir Robert Mansell ordered that all wood fuel be made available for the exclusive use of the Navy for the war ship fleet, leaving glassmakers to turn to coal deposits. Ironically, this secondary fuel source provided higher **firing** temperatures that, in turn, contributed to the quality of Ravenscroft's product. The coal furnaces required large drafts that led to enormous factories such as the Red House Glassworks in Wordsley, England. Its 90-ft. high conical-shaped flue stack was used for more than 125 years.

The shift from wood to coal also helped Sir Kenelm Digby, a leading British intellectual with a talent for scientific inquiry, to produce glass bottles at higher temperatures with a higher silica content giving a relatively inexpensive product with fewer blemishes and a resulting higher strength. The importance of these superior "English bottles" cannot be overemphasized (Figure 3.8). While bottles had been available from the first century AD, they had been used primarily for decanting and serving wine. Abundantly available, durable bottles allowed wine to be stored outside wooden casks. Storage within chemically inert glass allowed wine to be made with lower levels of sulfur, acidity, and tannins

Figure 3.7 A computer-generated image showing the random linkage of tetrahedra in the structure of silica glass, in contrast to the regular order of crystalline quartz in Figure 3.6. (Reproduced with kind permission of Sabyasachi Sen, Department of Materials Science and Engineering, University of California, Davis.)

Figure 3.8 A typical "English bottle" that evolved from the technological developments in glassmaking in the eighteenth century. (Reproduced with kind permission of the Corning Museum of Glass.)

compared with their necessity in cask storage. Once again, improvements in glass and wine went hand in hand.

Digby's bottles, dark green or brown in color, also had a "string rim" on the neck so a cork inserted for storage could be tied down with twine. Before the corkscrew, corks were left protruding from the bottle to facilitate easy removal. When bottle storage became possible, the upper class in England was able to buy wine in barrels and store them until ready to consume. At that time, the wine would be transferred to bottles carrying the family name or seal. The lead servant in the household would be charged with going to the cellar and carrying out this task. This "bottler" became known with time as the *butler* of the household.

The Methuen Treaty of 1703 between England and Portugal had significantly enhanced the availability of cork for bottle sealing, but, on August 24, 1795, the first English patent for a corkscrew was issued to a clergyman named Samuel Henshall effectively ushering in the modern era of fine wine appreciation. The bottle cork–corkscrew system continues to this day as the center of wine culture throughout the world. Figure 3.9 shows three contemporary

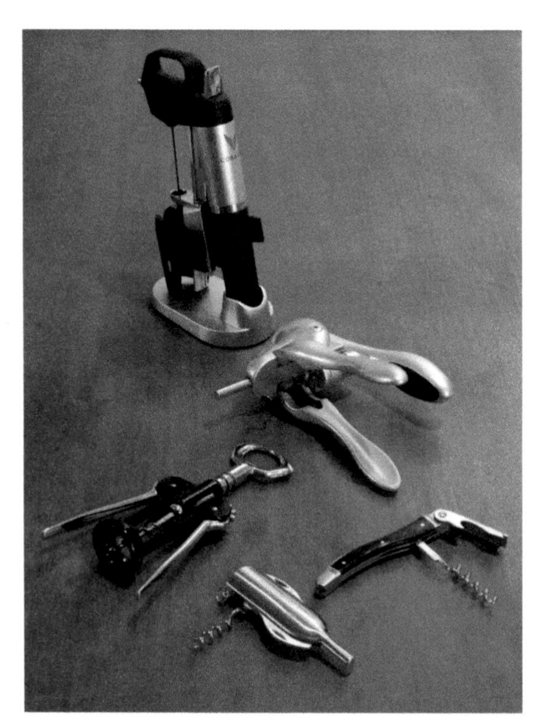

Figure 3.9 For more than two centuries, corkscrews such as the three designs in the foreground have been an integral part of wine appreciation. Behind them is a tool that uses a simple gear system to simplify and quicken the cork removal. In the background is a recent invention that eliminates the corkscrew altogether, extracting the wine through a needle under an atmosphere of inert argon gas.

corkscrew designs in the foreground that are little changed from the earliest models, along with a mechanized tool to simplify and quicken the process of cork removal, and finally, in the background, a recent device that avoids the corkscrew altogether by extracting the wine through a needle under an atmosphere of inert argon gas avoiding the remaining wine's exposure to air.

Among the many glass-and-wine developments of the eighteenth century was the evolution of bottles from the easily mouth-blown pear shape to more cylindrical shapes allowing the sideways stacking now routinely associated with wine storage. Progress in glassmaking in the nineteenth century shifted to the rising industrial power, America. In the 1820s, both Bakewell, Page & Bakewell in Pittsburgh and the Sandwich Glass Company in Sandwich, Massachusetts developed technologies to make glass using mechanical pressure reducing the need for human glassblowers. In 1881, American glassmaker Philip Arbogast made a creative suggestion that further laid the groundwork for the automation of glassmaking (Figure 3.10). He essentially turned the centuries-old technique of the glassblower upside down. Instead of making a bottle's neck and mouth last, he said why not have a machine make them first? This novel step would create a standard size object to hold onto while the rest of the bottle is formed. Unfortunately, he did not build the machine himself, rather he sold the patent and failed to reap the profits from what would prove to be the key to bottle-making automation.

The twentieth century opened with an extension of the use of automated air pressure and Arbogast's idea: a seminal contribution from the self-taught inventor Michael Owens. After many years of trial and error, Owens unveiled in 1903 the fully automated bottle-making system at the Libbey Glass plant in Toledo,

Figure 3.10 Philip Arbogast's brilliant contribution to glassmaking was to deliver a "gob" of glass to a mold, thereby forming the mouth and neck of the bottle first rather than last as done by traditional glassblowers. Arbogast's invention paved the way for the automation of bottle-making. This demonstration of his concept is in the Innovation Center of the Corning Museum of Glass. (Reproduced with kind permission of the Corning Museum of Glass.)

Ohio. This masterpiece of automation preceded Henry Ford's celebrated introduction of an automobile assembly line by a full decade and further reduced the role of the skilled craftsmanship of the glassblowers.

Michael Owens was, in fact, an unlikely "Edison of glass." He was not particularly mechanical by nature and could not read a blueprint. He had gone to work at the age of 10 shoveling coal at a glass factory in West Virginia, one of the leading glassmaking regions in the nineteenth-century America. By the still tender age of 15, he had moved up to being a glassblower on the shop floor. By the time his career had moved even further to the level of production supervisor at the Libbey Glass Company, he was convinced that a fully automated bottle machine was a legitimate goal. Owens' lack of mechanical aptitude was more than compensated for by his ability to visualize the glassmaking process. Owens had shared his vision in 1894 with company president Edward Libbey who guaranteed the financial backing. The unmechanical Mr. Owens enlisted the aid of the engineer Emil Bock with the simple admonition "Put it into iron, then we'll see if it works." With Bock's help, it did (Figure 3.11).

Owens' first successful prototype, completed in 1902, had been named pragmatically "Machine Number Four," and the revolutionary commercial machine was appropriately labeled "Machine A." Michael Owens was the model of the self-made man. His invention led to the formation of the Owens Bottle Machine Company in 1903, and in 1917, the Libbey-Owens Sheet Glass Company was formed, as sheet glass manufacturing was being automated in parallel with the progress in the container field. Owens' benefactor Edward Libbey was now his partner.

Libbey and Owens were not without rivals. Some of their competitors hired an engineer named Karl Peiler to find a way to feed the glass to the bottle machine in a more practical way. Owens' design essentially sucked up the heavy, molten glass into the machine. In 1915 after considerable hazardous experimentation, Peiler found an alternative: by controlling the temperature of the glass carefully, his design could drop a "gob" of glass of the correct size and shape into the machine's mold for final forming by a blast of air. His gob feeder is a central feature of modern IS (individual section) machines that produce most of the bottles in the world today. The term "section" of an IS machine refers to

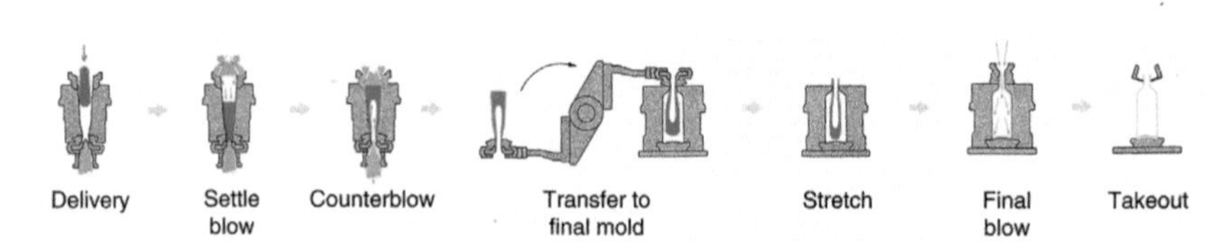

Figure 3.11 The complete automation of modern bottle-making is the descendant of Michael Owens' brilliant invention of 1903. The overall steps in the operation of the individual section (IS) machine are summarized in this illustration from the Innovation Center at the Corning Museum of Glass. (Reproduced with kind permission of the Corning Museum of Glass.)

the fact that each section is identical to the other with each section used to produce a separate bottle.

Owens' invention and its subsequent refinements were not just brilliant. They were big. By 1920, the Owens' bottle machine had grown to be a 30-ton behemoth with 10,000 parts, but the result was dramatic. It could produce more bottles in an hour than a team of human glassblowers could in a day.

Automation of glassmaking has not been limited to bottles. High-quality stemware can now be manufactured by machine. For example, Riedel Glass that had introduced varietal-specific stemware in the 1960s launched the Vinum brand in 1986 providing essentially the same range of shapes in a substantially more economical product line.

The chemically inert and pristine surface of glass has proven to be the perfect environment for wine, both within bottles that now allow wine to be routinely stored for decades and in fine "crystal" for the final enjoyment by connoisseurs. Glass is now synonymous with the appreciation of wine, both fine and not so fine throughout the world. The excellence of these contemporary bottles and stemware represents four millennia of technological progress from the first fusion of sand and other local ingredients in the Middle East to today's worldwide glass industry (Figure 3.12).

Figure 3.12 Glassware at this wine shop in Saint-Émilion, in the Bordeaux region of France, is the result of 4000 years of technological progress from the first human-made glass produced in Mesopotamia.

BIBLIOGRAPHY

Clinanti, Pino, *El Fuoco, Il Vetro, Il Vino*, Fondazione Banfi, Montalcino, Italy (1992).

Davila, Lilian, Subhash Risbud, and James Shackelford, "Interstitial Nanostructures in Engineered Silicates," *Ceramic Transactions*, **137**, 209–219 (2003).

Ellis, William S., *Glass*, Bard/Avon, New York (1998).

MacNeil, Karen, *The Wine Bible*, 2nd Edition, Workman, New York (2015).

Shackelford, James, *Introduction to Materials Science for Engineers*, 8th Edition, Pearson, Upper Saddle River, NJ (2015).

Smrček, Antonín, "Compositions of Industrial Glasses," in *Fiberglass and Glass Technology*, F. Wallenberger and P. Bingham, Eds., Springer, New York (2010).

Taber, George M., *To Cork or Not to Cork: Tradition, Science, and the Battle for the Wine Bottle*, Scribner, New York (2007).

Chapter 4

Modern Winemaking – A Role for Materials Other Than Glass and Ceramics

"God made only water, but man made wine."

Victor Hugo in *Les Contemplations* (1856)

Winemaking is a set of processes of astonishing complexity and variability. On the other hand, the ultimate goal of winemaking can be expressed by an astonishingly simple equation:

$$\text{Sugar} \rightarrow \text{Alcohol} \tag{4.1}$$

Equation 4.1 is the somewhat oversimplified story of fermentation, in which sugar in the wine grape is converted to alcohol in the presence of **yeasts**. To be more specific, we can write

$$C_6H_{12}O_6 \rightarrow 2\,C_2H_5OH + 2\,CO_2 \tag{4.2}$$

in which the glucose molecule represents sugar and the ethanol molecule is the form of alcohol found in wine and other alcoholic beverages. We also acknowledge in Equation 4.2 that an important by-product of the fermentation is the production of CO_2 gas. We can, in fact, thank Louis Pasteur, the great nineteenth century chemist and biologist (Figure 4.1), for this scientific foundation of winemaking after several millennia of largely trial and error efforts. Wine lovers can thank Professor Pasteur for more than just those other common elements of modern life: vaccinations and pasteurization. He made winemaking a science as well as an art.

The Glass of Wine: The Science, Technology, and Art of Glassware for Transporting and Enjoying Wine, First Edition. James F. Shackelford and Penelope L. Shackelford.

Figure 4.1 This memorial to Louis Pasteur on a wall in the Bordeaux region of France appropriately honors his contributions to the science of winemaking. His statement that "wine is the healthiest and most hygienic of drinks," has withstood the test of time.

As this book comes from the American Ceramic Society (ACerS) representing fields primarily within the physical sciences, those ACerS members and others from outside the fields of the biological sciences and organic chemistry might appreciate a brief glossary of terms that are less common in the ACerS literature but central to winemaking. These terms are provided in Table 4.1. (A full glossary for the entire book is given in Appendix B.)

Throughout this book, we look at the world of wine through a lens of the materials that make our wine experience possible, with glass being the primary "lens." In this chapter on winemaking, however, other materials take center stage. The two most important ones are dramatically different in nature – wood and steel. **Wood** is a natural material that has provided humankind with fuel and shelter throughout our history, but, for winemaking, a single wood has proven to be uniquely suited for containing the fermentation process as well for aging the wine. That wood is **oak**, and, while oak has served these central functions since the time of the Roman Empire, steel has become the other dominant material for winemaking within the past 50 years. Specifically, temperature-controlled **stainless steel** tanks have become a standard part of modern winemaking. This chapter will focus on these two materials central to the art and science of winemaking. In Chapter 5, we will find two ceramic materials that are effective alternatives to oak and stainless steel for fermentation vessels. Glass, however, is not entirely absent around the winery. A survey of the many uses of glass in various supporting roles is given next in Chapter 6.

So, what is it about *oak*? Why, of the more than 200,000 species of wood, has oak proven to be so effective for making wine? The answer lies first of all in its structure on the microscopic scale. In Chapter 1, we alluded to the various

Table 4.1 Glossary of Terms for Winemaking

Term	Definition
Cellulose	Compound with the formula $(C_6H_{10}O_5)_n$. Roughly half the composition of wood is cellulose in the form of fibers embedded in a matrix of lignin and hemicellulose
Dicotyledons	One of the two groups of flowering plants in which there are two (di-) embryonic leaves (cotyledons). The other group is the monocotyledons with one embryonic leaf. Oak, the ubiquitous material choice for wine barrels is in the dicotyledons group
Hemicellulose	A class of polymers similar to cellulose but with considerably lower molecular weights that, with lignin, forms the matrix in wood microstructures
Lignin	A class of phenol polymers that, with hemicellulose, forms the matrix in wood microstructures
Phenols	A class of compounds, the simplest of which is phenol with the formula C_6H_5OH, that serve as precursors to a wide variety of organic compounds. Phenols play an important role in the interaction of wooden barrels with wine
Tannins	Phenolic biomolecules with distinctive astringency (dry, puckery mouthfeel). Tannins play a significant role in the character of red wine
Vanillin	Phenolic compound with the formula $C_8H_8O_3$ that is the primary component of the extract of the vanilla bean. This organic compound can be transmitted to wine from oak barrels
Yeasts	Unicellular microorganisms of the fungus kingdom that convert carbohydrates to carbon dioxide and alcohol during fermentation

categories of engineered materials that include **metals**, **polymers** (**plastics**), and **ceramics** (and **glasses**). We find in Appendix A that covers the fundamentals of chemical bonding that the three types of **primary chemical bonds** (metallic, covalent, and ionic) provide the basis for those first three categories: **metallic bonding** defining the nature of metals, **covalent bonding** playing a central role in the nature of polymers, and **ionic bonding** giving us an appreciation of the nature of ceramics and glasses (while remembering Figure 3.5 to appreciate the structural difference between crystalline ceramics and noncrystalline glasses). There is however an additional, important category of engineered materials not primarily defined by chemical bonding but by microstructure, namely, **composites**. The most common example of this modern structural material is fiberglass with a microstructure composed of glass fibers a few micrometers in diameter embedded in a **matrix** of polymer. This human-made material has been used on a wide scale since World War II for myriad uses, including portable chairs, boat hulls, and piping. Its descendants include high-performance aerospace composites that generally involve more sophisticated polymer matrices and fibers other than glass, such as carbon, which leads us back to wood, a *natural* composite. In

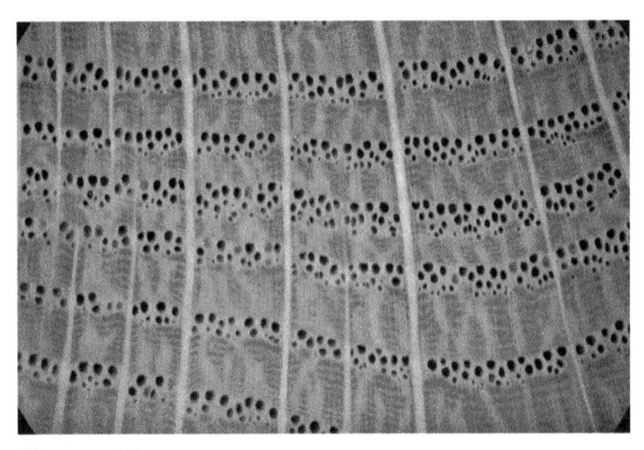

Figure 4.2 This optical microscope image reveals the complex microstructure of *Quercus robur*, a species of oak used in wine barrel making. The pores provide a route for water and nutrients to travel from the tree's roots up to its leaves. The porous, horizontal bands are annual growth rings that develop in the spring. About 6 years' growth is shown. (Reproduced with kind permission from Science Photo Library.)

the case of wood, **cellulose** fibers are embedded in a matrix of **lignin** and **hemicellulose**. Furthermore, the growth of this complex composite involves a porous microstructure in which the pores provide channels for water and nutrients to travel from the tree's roots up to its leaves. An image of the microstructure of oak is shown in Figure 4.2.

The critical component of oak's microstructure is that the complex combination of fibrous and matrix materials includes some open space. This porosity in oak, as seen in Figure 4.2, turns out to be ideal for winemaking, allowing a path for water and alcohol to evaporate outward while small amounts of oxygen can migrate into the barrel. We shall explore the complex relationship between wine and air (oxygen) in Chapter 14 in regard to the final step just prior to drinking in which some wines require aeration and/or decanting. Nonetheless, some minute amount of **oxidation** is a subtle part of the aging of those wines also, and the porosity of oak has proven to be most effective in this regard.

In spite of the enormous number of wood species, each wood is generally classified as either a softwood or a hardwood; softwoods are those coming from conifers such as pine, and hardwoods are those coming from broad-leaved **dicotyledons** such as oak. Most of the over 200,000 species are, like oak, classified as hardwoods. So, again, what is it about oak? Other common hardwoods have been considered, such as cherry, walnut, and chestnut, but the results are just not the same. One can argue that these other woods would "work" (providing water and alcohol evaporation and simultaneous oxygenation) but, at the same time, the results would be different. In a way, staying with oak has produced one less variable in the winemaking process. In fact, "staying with oak" is not without variability. There are approximately 600 species of oak, but four are most commonly used in making barrels for winemaking. The most prized (and generally most expensive) are two species associated with French winemaking:

Figure 4.3 These gravity-fed wooden barrels at Château Brane-Cantenac in the Bordeaux region of France are a beautiful expression of the central use of wood in traditional winemaking. (Reproduced with kind permission of Alexia Defranoux, Château Brane-Cantenac.)

Quercus robur and *Quercus sessiliflora* (or *Quercus petraea*). A popular alternative is American oak or *Quercus alba* (used widely not just in America but in other countries, especially Spain). Another economical alternative is Hungarian oak or *Quercus frainetto*. The greater porosity of French oak leads to greater oxygen access and correspondingly greater "softening" of the wine during barrel aging compared to the denser American oak. (The "silky" character of an expensive cabernet sauvignon is an ideal result.) The rise in popularity of Hungarian oak is coming from the fact that it is relatively economical and closer in character to French oak than American. Figure 4.3 shows an opulent use of wood in the gravity barrel room in a Bordeaux winery. The use of gravity to move the wine into and out of aging barrels is preferred by some winemakers to minimize the use of pumps and the subsequently gentler handling of the wine.

While we use microstructural porosity as our first basis for choosing oak, chemistry is the second important aspect. Specifically, oak contains a number of organic chemical compounds central to winemaking, especially **phenols**. Among the most distinctive results of these phenols migrating into the wine is a vanilla taste due to the presence of the phenolic compound **vanillin**. American oak can also provide a coconut flavor in addition to vanilla. Equally important is oak's source of phenols called **tannins**. Their effect on red wines is not unlike the bitterness and astringency they provide in black tea, leading to a dry and puckery feel in the mouth. Tannins also play a critical role in the long-term storage of red wines, as will be discussed later in this chapter. Just as European oaks provide more oxygen porosity than the American counterpart, European oaks also provide more tannin extraction. These biochemical effects provided by a barrel are, in any case, most profound on its first use. The effects are substantially less on second use, and, after four uses, little if any flavoring is produced by what is now termed a "neutral" barrel. The slight oxygenation provided by the microstructural porosity is nonetheless still available, and these neutral barrels can be used for decades longer.

Next, we move from the natural world of wood to the engineered world of stainless steel. Stainless steel is one of the iconic examples of an engineered material that has evolved out of the metallurgical wing of the field of materials science and engineering. **Steel** is an iron-based alloy that contains up to about 2 wt% carbon, along with other alloying additions in some cases. Steel becomes "stainless" by the addition of certain alloying elements that give the iron resistance to chemical reaction with its environment. (*Stainless* is, in fact, a marketing term as some chemical reaction can still occur and substantial corrosion is even possible in certain aggressive environments.) Chromium (Cr) is the primary alloy addition that provides this chemical resistance, with at least 4 wt% needed and amounts greater than 10 wt% being common in commercial products. The specific mechanism of protection is the formation of an (Fe, Cr) oxide coating on the surface of the alloy that serves as a barrier to chemical reaction with atmospheric oxygen. Without the addition of chromium, iron can form iron hydroxide, better known as rust, when atmospheric oxygen reacts with surface moisture and the unalloyed iron.

The ability of chromium additions to improve iron's resistance to acid attack was discovered in the early nineteenth century and, within a hundred years, low-carbon chromium supplies were available to produce alloys comparable to modern stainless steel compositions. One of the early leaders in the development of this new material was the Krupp Iron Works of Germany. In 1912, Krupp filed for a patent on a chrome–nickel–molybdenum stainless steel with the additional alloying elements providing improved acid resistance over simple chromium steel. The use of stainless steel tanks for wine production would not become practical, however, until the 1950s when rotary compressors capable of refrigerating 95,000 l (25,000 gallon) tanks in the winery become available. By the end of the 1960s, temperature-controlled stainless steel tanks, such as those shown in Figure 4.4, had become commonplace.

The two main stainless steel alloys used for wine tanks are the 304 and 306 types, where the numerical codes specify particular compositional ranges for the various alloying elements. For these materials that are widely used in a broad range of industries, the chromium composition is around 16–20 wt% and the nickel composition is around 8–14 wt%. The 304 material is commonly referred to as "18-8 stainless" based on typical chromium/nickel levels. Both types are referred to as "austenitic stainless" alloys, an historical reference to their common crystal structures (face-centered cubic for those familiar with crystallography). Both alloys can have up to 2 wt% manganese and 1 wt% silicon additions as well. The 316 alloy also has 2–3 wt% molybdenum, providing a greater resistance to SO_2 and various acids found in the fermentation tanks. In general, the hostile environment supplied by the fermentation process can challenge the "stainless" description, and special care is required to prevent undermining

Figure 4.4 Temperature-controlled stainless steel tanks are widely used alternatives to the more traditional oak barrels for winemaking. These contemporary examples are in the Teaching Winery of the Robert Mondavi Institute (RMI) for Wine and Food Science in the University of California, Davis campus. (Reproduced with kind permission of Andrew L. Waterhouse, Robert Mondavi Institute for Wine and Food Science.)

the corrosion protection by abrasive cleaning methods or cleaning solutions with excessive levels of chlorine.

MAKING RED WINE

So, now let us follow the use of oak barrels and stainless steel tanks in making a red wine, specifically a cabernet sauvignon (Figure 4.5). The sweet juice of the cabernet sauvignon grape is, in fact, colorless (as is the juice of most grapes whether being used to produce red or white wine). The redness of the wine is the result of allowing the fermentation to take place in the presence of the grape skins. Like fresh barrels, grape skins contain tannins, so allowing fermentation to take place in their presence, the skins are providing tannins as well as color. The heat evolution that accompanies fermentation as well as the leaching ability of the alcohol being produced facilitates this transfer of skin tannins to the wine. As a result, red wines are more tannic than whites that are fermented without significant skin contact. As a practical matter, the grape stems also contain tannin, and many winemakers will choose to put the grape clusters into a

Figure 4.5 Great wine requires great grapes, such as these from a vineyard in the Bordeaux region of France.

crusher–destemmer machine to remove those stems to avoid an excessive amount of tannin in the final product. (For a less tannic variety, such as pinot noir, the winemaker might leave the stems in place during fermentation to provide an additional boost in tannin level.) Fermentation is often done in stainless steel allowing easier temperature control and clean up than in wooden barrels or vats, the common practice until the 1960s. The temperature control available with stainless steel tanks is very helpful in avoiding excess heat. Even reaching 30 °C (86 °F) can degrade delicate aromas and flavors.

Returning to the discussion relative to Equation 4.2, we are reminded of the central role of yeasts in winemaking. So-called native yeasts are present in vineyard soils where they attach to the grapes and are also present on surfaces and in the air of the winery itself. Some traditionalists like to carry out the fermentation with these native yeasts, but the process can be slow and subject to challenges such as attack by spoilage bacteria. As a result, many winemakers today prefer using cultured yeasts to facilitate a more rapid fermentation with greater control of the overall process. This decision is just one of many that the winemaker must make throughout the winemaking process, not unlike the work of a conductor preparing for a musical concert (Figure 4.6).

Fermentation is accompanied by the evolution of substantial heat and, as seen in Equation 4.2, a substantial amount of CO_2 gas. Left unattended, the grape skins are pushed up from the fermenting mass by the CO_2 gas to form a

Figure 4.6 Penelope enjoys a tour of the barrel room at Château Léoville Poyferré with winemaker Isabelle Davin. Ms. Davin was one of the first female winemakers in the tradition-bound Bordeaux region of France. (Reproduced with kind permission of Florence Cuvelier, Château Léoville Poyferré.)

cap over the liquid. In order to ensure adequate transfer of color, tannins, and associated flavor and aroma, the cap is pushed back into the liquid regularly. While stomping about in the fermentation tank in bare feet is a popular image, most wineries today use wooden paddles and similar tools to do the job. In some areas, such as the Duoro Valley of Portugal where Port wines are produced, foot stomping is still quite popular. This tradition is not a sign of primitivism, as the human foot is mechanically ideal for the task of crushing grapes. Nonetheless, many people prefer another way to earn a wage and welcome mechanical substitutes.

The conversion of the grape's sugar to the wine's alcohol is typically complete after a period of 5–14 days in the fermentation tank. The result is a so-called "dry wine" in which the term "dry" means "not sweet," even though the wine is obviously a wet liquid. Most table wines are considered dry wines, with many books on wine appreciation emphasizing the difference between the terms "sweet" and "fruity." A wine with no residual sugar can have a strong fruitlike character and hence is termed *fruity* even though no sweetness is present. Nonetheless, we tend to associate fruit with sweetness long before we begin appreciating wine, which leads to a tendency for the two terms to be used (incorrectly) interchangeably. On the other hand, fermentation, even in so-called dry wines, does not always go to absolute completion. While there is no widely accepted standard for the dry-to-sweet levels for table wines, the dry label can still be generally applied to a wine with as much as 0.5% residual sugar.

The alcohol level of the finished dry wine can range from a low of 10% to a high of 16%, depending on the type of grape and its ripeness at harvest. The upper limit is essentially self-policing by the wine as the yeasts essential to the process die of alcohol poisoning beyond that concentration! During the latter part of the yeast-driven fermentation process of red wines such as cabernet sauvignon and/or extending beyond it, a separate process is used, namely, malolactic fermentation in which *Oenococcus oeni* bacteria drive the fermentation converting the somewhat tart malic acid in wine to the softer lactic acid found in milk. An additional by-product is also dairy related: the diacetyl molecule found in butter. While an excessive buttery quality is to be avoided, malolactic fermentation provides needed softness to the red wine character as well as microbial stability.

Once fermentation is complete, the wine is drained off the skins and is ready for aging. (Additional, somewhat lower quality wine is available by gently pressing the remaining skins to release additional juice.) Premium wines are then aged in oak barrels for a few months up to a few years to provide the effects described earlier in terms of the nature of oak wood. During barrel aging, the wine is periodically "racked," that is, clear wine is drawn off after solids settle to the bottom of the barrel. The racking process also provides some aeration that is a natural part of the maturation of the wine. "Fining" is another process that is sometimes used in which protein coagulants, such as egg whites, are used to clarify the wine by removing minute solids such as unstable protein suspensions and excess tannins. Fining is distinct from "filtering," which has become a somewhat more controversial technique. A typical filter is composed of cellulose fibers and is intended to clarify the wine (removing suspended particles) without removing flavors or aromas. The controversy has arisen after many wine critics have felt that filtering is used too widely and has sometimes resulted in relatively bland wines in comparison to unfiltered counterparts.

Following barrel aging, fining, and filtering (or the absence thereof in the case of *vin ordinaire*), aging continues in the bottle. From this point, the pristine surface of the glass bottle causes the additional aging to be a process internal to the wine itself without the possibility of water and alcohol evaporation, as was the case in the barrel. The subtle possibilities of oxygen access through a cork closure as opposed to other options is discussed in Chapter 13. In any case, aging a highly prized and strong tannic red wine such as those from the great Châteaux of Bordeaux can go on for decades (Figure 4.7). During extended bottle aging, the presence of harsh tannins diminishes producing a softer mouthfeel, and eventually the color will tend to change from red to brown. This production of brown or "tawny" color is a distinctive characteristic of decades-old tawny Ports that achieve their color by spending most of those decades in barrel, prior to bottling, as will be discussed toward the end of this chapter. All of these changes are happening as a result of the chemical reactions of phenolic compounds in the wine, with some of these compounds binding together to form sediment that should be carefully decanted prior to serving. Excessive aging is indicated by a significant dissipation of the fruit flavors of the wine and the dominance of its acidity.

Figure 4.7 Bordeaux wines aged for more than a century at Château Cos d'Estournel. (Reproduced with kind permission of Château Cos d'Estournel.)

MAKING WHITE WINE

Now, let us turn to producing a white wine, such as chardonnay, also using oak barrels and stainless steel tanks. Much of the overall process is similar starting with Equations 4.1 and 4.2, so let us focus on differences. First and foremost and as we noted at the outset of discussing red winemaking, the skins are removed *before* fermentation begins when making a white wine. The relatively gentle juice press may also be preceded by passage through a crusher–destemmer machine. After some time in a settling tank to allow minute pieces of grape pulp to settle out, the clear juice is ready for fermentation. While the winemaker's decision to use either native or cultured yeasts is similar to that he or she must face in making a red wine, the use of a stainless steel tank is less arbitrary. The ability to maintain relatively cool fermentation temperatures in these modern vessels has provided a worldwide revolution in white wine production. Until they became available in the 1960s, white wines of great freshness and delicacy were predominantly produced in the cool climates of Germany and Northern France. Today, such conditions can be artificially produced in once hostile locations from the warm climes of Italy to California. Stainless steel tanks for making premium quality white wines in Sicily are seen in Figure 4.8.

While we have extolled the virtues of stainless steel tanks for improving the quality of white wine production around the world, we must also acknowledge that producing chardonnay in oak barrels in California has continued to be popular, with the toasty, vanillin character a basis of that popularity. Furthermore, while malolactic fermentation was a commonplace practice in making red wine, its use in white winemaking is another important decision for the winemaker and far from universal. The popularity of "buttery chardonnays" from California has led to something of a backlash, with critics likening the result to "theater popcorn," and some winemakers avoiding that character emphasizing that their

Figure 4.8 Stainless steel tanks have allowed high quality white wines to be produced worldwide. As an example, first-rate Etna Bianco is produced at the Pietradolce Winery on the Italian island of Sicily. (Reproduced with kind permission of Giuseppe Parlavecchio, Pietradolce.)

chardonnay "has seen only stainless steel!" Although there is a spectrum of choices for chardonnay, crisp wines like rieslings with their high level of acidity are inappropriate candidates for malolactic fermentation.

An additional source of "richness" for wines, such as chardonnay, is to allow the wine to remain for several months after fermentation is complete in the presence of the spent yeast cells that have settled to the bottom of the tank. The dead cells are called "lees," and this technique is called spending time "on the lees," or by the equivalent French term *sur lie*. The interaction between the wine and these cells can add texture and complexity to the wine, effects that can be enhanced by occasionally stirring the lees.

A common next step for most white wines is *cold stabilization* in which the wine is chilled to a temperature just above freezing for several days in order to precipitate out tartaric acid. This is a largely cosmetic exercise in order to avoid the presence of these tasteless crystals in the final product.

While barrel aging of fine red wines is routine, the aging of white wines in oak barrels is not. Again, chardonnay is one such "full-fruit" white wine that is frequently barrel aged, but, as with the careful control of the extent of malolactic fermentation, the amount of time in oak barrels must be controlled with care to avoid excessive "wood character," especially vanillin. And, finally, the winemaker, as with red wine, can choose to clarify the wine further by fining or filtering before bottling. In any case, further aging is possible in the bottle, although, for white wines, the aging process is not a story of diminishing harsh tannins as it was with fine red wines. The aging of fine white wines is related more to the level of acidity. High acid varietals that contain some sweetness are capable of decades long storage. Prized German rieslings are prime examples. Chardonnays, on the other hand, tend to be aged no more than 5 or 6 years in bottle.

MAKING ROSÉ WINE

Discussing the similarities and differences in making red versus white wines begs a rather obvious question: how does one make rosé wines? The most obvious approach is to follow the technique for making red wine but to limit skin contact with the juice to a relatively brief period of one to three days. An alternative used in some parts of France is the *Saignee* (or "bleeding") method in which some juice is bled off the so-called *must* or soupy combination of crushed grapes, skins, pulp, seeds, and so on. In this case, the rosé wine is a by-product of the winemaker's effort to make a more intense red wine in the remaining must that has had some of its juice bled off. Some of the most prized rosés come from the area around the fishing village of Bandol in the South of France (Figure 4.9). Finally, a rather inelegant approach is to simply blend some red and white wines.

Figure 4.9 Some of the most prized rosés in the world come from the Bandol area of Provence in the South of France, where this poster hangs in the region's most famous winery, Domaine Tempier. (Reproduced with kind permission of Veronique Peyraud Rougeot, Domaine Tempier.)

MAKING SPARKLING WINE

There is perhaps no more elegant expression of wine culture than Champagne. Frequently served prior to a fine meal, it is also the "gold standard" for enriching life's most special events from weddings to baptisms to anniversaries. Perhaps the only downside to this elegant experience is the price to be paid for the more famous expressions of the sparkling winemaker's skills from the leading Champagne houses of France. Fortunately, there are other delightful sparkling wines that tend to be much more friendly to the pocketbook from less prestigious Champagne brands to a large number of cava wines from the Catalonia region of Spain and Prosecco wines from the Veneto region of Northern Italy (Figure 4.10).

In general, making these sparkling wines begins by making a still wine followed by a second fermentation to produce the characteristic bubbles. The best Champagnes are generally made from either chardonnay (white) or pinot noir and pinot meunier (red) grapes. The corresponding terms that can appear on labels are *blanc de blancs* for a golden/white sparkler made entirely from chardonnay and *blanc de noirs* for a slightly pink/golden product made entirely from red grapes. In fact, *blanc de noirs* wines represent a style made largely outside of Champagne, especially in California. On the other hand, rosé Champagnes are highly prized being made by either the *saignée* (or "bleeding") method described earlier or simply adding some still pinot noir wine to a bottle just before the second fermentation.

Figure 4.10 High-quality Prosecco wines like these from the Bele Casel winery in the Veneto region of Northern Italy provide economical alternatives to the iconic sparkling wines from the Champagne region of France. (Reproduced with kind permission of Paola Ferraro, Società Agricola Bele Casel.)

In Champagne, several still wines will be made initially from a range of vineyards. These so-called base wines will later be blended to the taste of an experienced senior winemaker in preparation for the important bubble-producing second fermentation. The initial fermentation is frequently done in stainless steel tanks to take advantage of their temperature control, but some traditionalists still use neutral oak barrels. A large number of such base wines are typically used in the blend and are termed the *assemblage* in French. Unlike the case for most table wines, the tradition in Champagne is to focus on nonvintage blends to optimize quality rather than depending on specific vintages that can vary widely in quality.

The all-important second fermentation is created by adding some yeasts and a mixture of sugar and wine (*liqueur de tirage* in French) to the nonvintage blend. This whole system is bottled and capped. The second fermentation creates some more alcohol and, more importantly, additional CO_2 gas that is held in solution within the closed system (the capped glass bottle).

The regulated Champagne method (*méthode champenoise*) requires aging of these bottles for at least 15 months, with as much as 3 years or more of aging common. This step in Champagne making is a splendid example of the *sur lie* technique in which the ultimate richness of the wine is the result of this extended time on the lees. A by-product of this step is also a dense cloud of yeast cells. To clarify the wine, the bottles are turned upside down and ritualistically rotated numerous times until the yeast cells have collected in the neck of the bottles (a process that, as one might expect, has been automated over the years). To rid the wine of this yeasty sludge, the neck is frozen in a glycol solution that allows the bottle to be quickly turned upright and uncapped, causing the frozen batch of yeasts to shoot out of the bottle. This process known as *degorgement* is accompanied by the loss of about 5 mm (1/4 inch) of space at the top of the bottle. This gap is filled by a combination of wine and sugar, with the amount of sugar (known as the *dosage*) determining the ultimate level of sweetness of the final wine. At this point, the bottles are closed with the traditional flared bottom cork with a nearly spherical top, a design that facilitates the maintenance of the substantial internal pressure (about 5–6 atmospheres) and ultimately the ease of opening (with a pop!). While sweetness levels for table wines are not standardized, Champagne and similar sparkling wines have a "sweetness vocabulary" as summarized in Table 4.2. The fact that the various brut Champagnes are not as sweet as an "extra-dry" is, of course, a potential source of confusion among consumers. As a practical matter, most of the contemporary Champagne market consists of the brut level of sweetness.

Among the most popular and economical alternatives to Champagne are the Proseccos of Northern Italy made primarily from the glera grape. These widely available wines are a subset of the *spumantes*, the general name for any sparkling wine from Italy. In the case of Prosecco, they are regulated by a *Denominazione di Origine Controllata e Garantita (* DOCG*)*, the highest classification for an Italian wine in which the production methods are controlled and the

Table 4.2 Glossary of the Sweetness of Champagne and Similar Sparkling Wines

Term	Definition
Brut nature	<0.3% residual sugar
Extra brut	0–0.6% residual sugar
Brut	0–1.2% residual sugar
Extra dry	1.2–1.7% residual sugar
Sec	1.7–3.2% residual sugar
Demi-sec	3.2–5.0% residual sugar
Sweet	>5.0% residual sugar

quality is guaranteed. The labeling terminology for Proseccos is comparable to that for Champagne (Table 4.2). A primary distinction of Prosecco is that the secondary fermentation is generally produced by the alternative Charmat process using pressurized tanks rather than bottles.

While inexpensive versions of Prosecco are produced in millions of bottles per year by large wineries, more expensive and higher quality examples are increasingly available from small producers who are focusing on specific vintages from individual towns. Nonetheless, these premium Proseccos still tend to be a bargain in comparison to many Champagnes.

MAKING FORTIFIED WINE

We close our discussion of the main examples of winemaking appropriately enough with a discussion of how the wines enjoyed at the end of a meal are made; specifically the fortified wines. These powerful wines that combine substantial levels of alcohol with substantial sweetness pair well with desserts, and the classic example of a wine that combines the alcohol and sweetness with elegant flavors and aromas is Port from the Duoro Valley of Northern Portugal (Figure 4.11). The mouth of the Duoro River emerges at the city of Porto, Portugal's equivalent of Bordeaux, a city whose history and personality center around wine.

We could easily devote an entire chapter to the making of Port wine but will limit our discussion to three principle paths that lead to three of the most popular (of the ten total) styles of Port. First is vintage Port. For most wines, the word "vintage" simply indicates the year in which the wine was grown and harvested. Wine connoisseurs typically acknowledge some great vintages and some less so and still others that disappoint, but nonetheless each year is another vintage. For Port, however, only the most excellent years are declared a vintage, a designation that must be agreed to by at least half of the shippers of the wine and further agreed to by the powerful Port Wine Institute that oversees the industry. Such agreed-to vintages have been declared for only one in 4 years on average over the past century! And so, how is this most excellent expression of Port wine

Figure 4.11 Tawny Port is the very epitome of aged wine, as it rests for multiples of decades in large barrels, such as these at the Cálem Port house in Porto, Portugal. (Reproduced with kind permission of Porto Cálem.)

produced? We already noted that the traditional ritual of crushing grapes under foot has survived in the Duoro Valley vineyards more than almost any other winemaking region in the world. We should also note at this point that the primary grapes used to make Port are indigenous to this region, namely, Touriga Nacional, Touriga Franca, Touriga Roriz, Tinto Cão, and Tinta Barroca. As with all Ports, the combination of both high alcohol level and substantial residual sugar is achieved by adding a high (77%) alcohol clear brandy to the grape blend well before fermentation is complete. The result is that, as noted earlier, the high alcohol level kills the yeasts terminating further fermentation and producing a wine with substantial residual sugar (about 7%) and a final alcohol level of about 20%. Beyond this point, different aging strategies lead to the distinctly different styles of Port. For vintage Port, the high quality and highly prized wine is initially aged 2 years in barrel and then transferred to bottle for extended aging, with a decade being typical and the potential for many more decades of aging. Combined with the fact that Port grapes have thick skins loaded with tannin and the wine is not fined or filtered, substantial sediment is to be expected requiring decanting. (See Chapter 14 for a discussion on decanting and the special considerations regarding the aeration of Port wines.)

An economical alternative to a vintage Port is the late-bottled vintage (LBV) Port that is aged longer in barrel (4–6 years) but considerably less in bottle. An important distinction for LBV Ports is that "vintage" in this case is defined in a similar way to that for traditional table wines, that is, they can be bottled in any year, not just the most excellent ones. Nonetheless, in those excellent years, the LBV is also coming from the same high-quality fruit and the wine can be quite competitive with the considerably more expensive vintage Port.

Finally, a very different aging strategy leads to the production of aged tawny Port. In this case, the fortified wine is aged for literally decades in barrel, leading to the characteristic brown to brown-orange or tawny color. The label reflects the time frame declaring "10 Years," "20 Years," "30 Years," and beyond. As a practical matter, the declared age, such as 20 Years, is a rough average of the time that a blend of wines have spent in various barrels.

With an overview in hand of how table wines are made (red, white, and rosé) and the special effort required to make sparkling and fortified wines, all by using oak barrels and stainless steel tanks, we will return to our menu of engineered materials in the next chapter to appreciate two important options from the ceramic industry: concrete tanks (and "eggs") and ceramic amphorae.

BIBLIOGRAPHY

Amerine, Maynard and Vernon Singleton, *Wine: An Introduction*, 2nd Edition, University of California Press, Berkeley (1977).

Boulton, Roger, Vernon Singleton, Linda Bisson, and Ralph Kunkee, *Principles and Practices of Winemaking*, Springer, New York (1999).

MacNeil, Karen, *The Wine Bible*, 2nd Edition, Workman, New York (2015).

Shackelford, James, *Introduction to Materials Science for Engineers*, 6th Edition, Pearson Prentice Hall, Upper Saddle River, NJ (2005).

Chapter 5

Ceramics Around the Winery – Alternatives to Oak and Stainless Steel

"Other than goatskin bags, the earliest vessels for transporting wine in the ancient world were amphorae, terra-cotta jars with two looped handles and, usually, a pointed base."

Karen MacNeil, *The Wine Bible*, 2nd Edition

In her exhaustive and highly readable coverage of the world of wine, *The Wine Bible*, Karen MacNeil gives ceramics their proper place at the beginning of wine culture. Her "terra-cotta" jars are among the earliest ceramic materials from the ancient world with terra cotta simply meaning "baked earth," making these ancient ceramics a classic example of **earthenware**. Such low-fired ceramics are distinct from more modern ceramics that could be produced at higher temperatures (above 1200 °C) at which the claylike raw materials could be *vitrified* or transformed from the crystalline minerals to having at least some noncrystalline glassy phase. A by-product of this vitrification that occurs in making high-fired ceramics such as porcelain and stoneware is a relatively dense material in contrast to the characteristically porous earthenware. The blessing and curse associated with such porosity will occupy us throughout this book as we frequently return to the complex relationship between wine and oxygen. In this chapter, we focus not on *glasses* but on the chemically similar and structurally distinct *ceramics* (another reminder of the fundamental definitions provided by Figure 3.5). While dwelling on the past, we can acknowledge that

The Glass of Wine: The Science, Technology, and Art of Glassware for Transporting and Enjoying Wine, First Edition. James F. Shackelford and Penelope L. Shackelford.

the word "ceramic" comes from the Greek synonym for earthenware or baked earth, *keramos*.

While glass is not the central material used in winemaking, two of its ceramic cousins are sometimes used in place of oak and stainless steel. These cousins are, like glass, inorganic and nonmetallic in nature. Concrete is a popular alternative to stainless steel for holding tanks, generally in rectangular shapes in contrast to the cylindrical shapes for stainless steel tanks, and is sometimes configured as large egg-shaped containers. Also, several modern winemakers wishing to emulate traditional and even ancient winemaking techniques use ceramic amphorae similar to those introduced in Chapter 2 (Figure 2.2). And so our primary menu of materials for winemaking involves all four of the candidates in both the previous and current chapters: oak, stainless steel, concrete, and earthenware ceramic. As emphasized in the previous chapter, a survey of the many uses of glass in various supporting roles around the winery is given in Chapter 6.

With our understanding from Chapter 4 of how table wines (red, white, and rosé) as well as sparkling and fortified wines are made using oak barrels and stainless steel tanks, we can return to our menu of engineered materials to appreciate two important options from the ceramic industry: concrete tanks (and "eggs") and ceramic amphorae.

Concrete is such a widely used construction material that the concrete industry has an identity somewhat separate from the broad field of ceramics that also includes pottery, traditional whitewares (plates and dishes), sanitary ware (toilets), and high-tech materials such as high-purity oxides (aluminum oxide, zirconium oxide, etc.). Nonetheless, concrete is an appropriate category of the ceramic family of materials. One can also make the case that, like wood, concrete can be considered a composite material in that it is composed of two primary components: common aggregate (both coarse (gravel) and fine (sand)) held together in a calcium aluminosilicate (**cement**) matrix. In the case of concrete however, its microstructure is at a larger scale than composites such as fiberglass. For concrete, the coarse aggregate (common gravel) has a size range of 10–150 mm diameter and the fine aggregate (common sand) has a size range of 0.1–5 mm diameter. These millimeter-sized aggregate particles are considerably larger (by about three orders of magnitude) than the micrometer-scale fibers associated with common fibrous composites such as wood and fiberglass. The "glue" that holds the coarse and fine aggregates together in a tightly packed structure in concrete is a complex combination of ceramic oxides of primarily calcium silicates and calcium aluminates, with the overall cementitious material resembling limestone from the Isle of Portland in England, hence the commercial name *portland cement*. A cross section of a common concrete structure is shown in Figure 5.1.

In many parts of the winemaking world, concrete tanks are popular alternatives to stainless steel with their inherent thermal insulating characteristics and

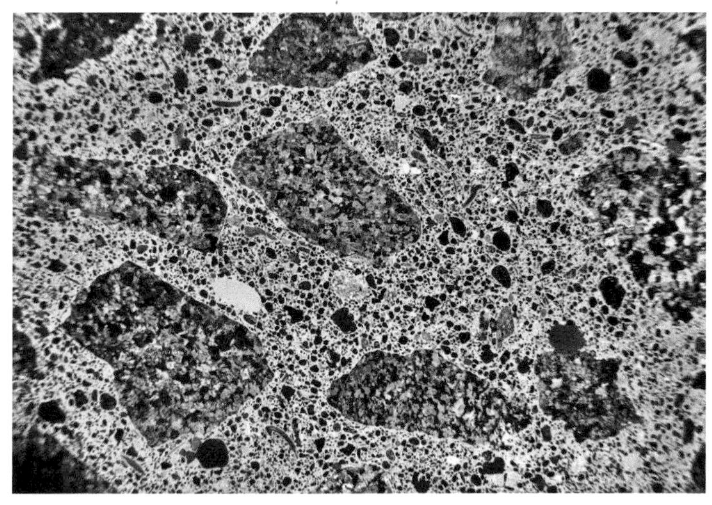

Figure 5.1 This polished cross section of a structural concrete can be described both as a ceramic (based on its silicate chemistry) and as a composite (based on the combination of gravel and sand aggregates embedded in a matrix of portland cement). (Reproduced with kind permission from Science Photo Library.)

Figure 5.2 These rectangular-shaped concrete tanks at Château Brane-Cantenac in the Bordeaux region of France are framed in stainless steel. Nonetheless, the wine sees only the concrete surface as it undergoes the fermentation process. (Reproduced with kind permission of Château Brane-Cantenac.)

ease of cleaning, as well as a natural level of porosity not unlike that for wooden vats (Figure 5.2). Also popular are egg-shaped concrete tanks that are easy to fabricate and are of intermediate size (Figure 5.3). Some winemakers feel that the egg shape creates a natural vortex during the fermentation process, reducing the need to stir the lees (dead yeast cells), as described in Chapter 4 while discussing the making of both still white and sparkling wines.

As winemaking has evolved from the scientific foundation provided by Louis Pasteur in the nineteenth century on through advancements in materials technology such as the temperature-controlled stainless steel tank, an ironic movement has evolved in more recent decades – the return to "natural" winemaking techniques. Contemporary winemaker Frank Cornelissen, a Belgian wine agent turned winemaker in Sicily, has become a leader in the movement to produce wines in a traditional or natural way. Cornelissen's philosophy is that earlier winemaking methods provided a more direct expression of the true nature of the grape's fermentation, and he purposefully tries to avoid many of the interventions provided by modern technological advances. Central to his approach is the use of ceramic amphorae, not unlike those used by the ancients that we

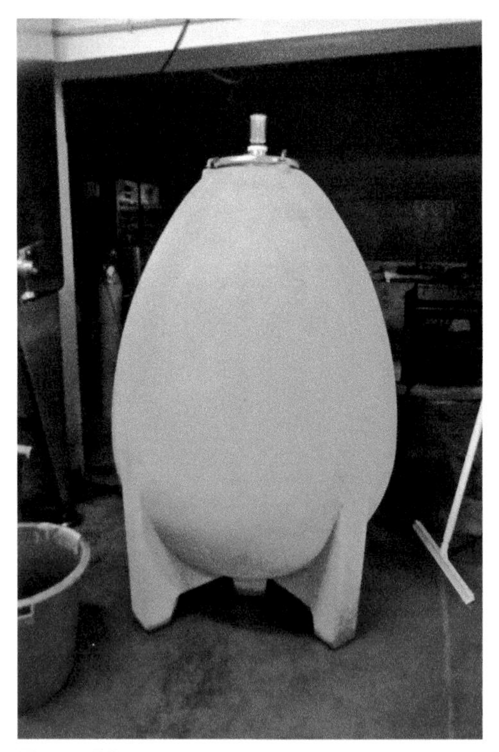

Figure 5.3 A smaller alternative to the concrete tanks shown in Figure 5.2 is the cast concrete egg vessel at the Domaine Tempier Winery in Bandol, France. (Reproduced with kind permission of Veronique Peyraud Rougeot, Domaine Tempier.)

Figure 5.4 These ceramic amphorae at the Casa Belfi Winery in Northern Italy allow a return to more traditional, even ancient, winemaking practices. (Reproduced with kind permission of Casa Belfi.)

Figure 5.5 Outside the Casa Belfi Winery, ceramic amphorae allow large-scale wine storage. (Reproduced with kind permission of Casa Belfi.)

covered in Chapter 2. These ceramics are akin to traditional **clay**-based pottery and are typified by the mineral **kaolinite**, a hydrated aluminosilicate with the formula $Al_2Si_2O_5(OH)_4$. The slightly porous microstructure of these clay-based ceramics can, like concrete, provide oxygen transport without contributing any tannins or coloration. Figure 5.4 shows a beautiful example of such amphorae produced in Spain that are being used by the Casa Belfi winery in Northern Italy to produce an organic Prosecco sparkling wine. While these amphorae are roughly the size of a typical oak barrel, Figure 5.5 shows Penelope standing next to exceptionally large amphorae outside the winery that allow large-scale wine storage.

While acknowledging wood and steel as the primary materials for winemaking and the significant role for concrete and ceramic alternatives, we can now move on to the wide spectrum of uses for glass around the modern winery, prior to it becoming the dominant material for wine storage, shipping, and consumption.

BIBLIOGRAPHY

Amerine, Maynard and Vernon Singleton, *Wine: An Introduction*, 2nd Edition, University of California Press, Berkeley (1977).

Asimov, Eric, "The Evolution of a Natural Winemaker," *New York Times*, August 25, 2016.

Boulton, Roger, Vernon Singleton, Linda Bisson, and Ralph Kunkee, *Principles and Practices of Winemaking*, Springer, New York (1999).

MacNeil, Karen, *The Wine Bible*, 2nd Edition, Workman, New York (2015).

Shackelford, James, *Introduction to Materials Science for Engineers*, 6th Edition, Pearson Prentice Hall, Upper Saddle River, NJ (2005).

Chapter 6

Glass Around the Winery – From Barrel to Lab

"I'm afraid that – not necessarily deliberately, but consistently – I've made a kind of laboratory out of my life, where I mix the stuff in the test tubes . . ."

Jay McInerney, writer and wine critic as quoted in *The Guardian* September 10, 2000.

The first three chapters made, we would hope, a convincing case for what a friendly surface glass provides to wine throughout its storage, shipping, and consumption. The last two chapters, however, focused on other materials central to the initial winemaking process. There are nonetheless myriad uses for glass around the winery during the production of wine. While glass may play ancillary roles compared to oak, stainless steel, concrete, and ceramics discussed in Chapters 4 and 5, its contributions are critical. In surveying a few of these roles, we will unveil a gallery of scientific glassware from classic analytical chemistry techniques to the inner workings of some of the most contemporary scientific instruments. Along the way, we will be introduced to factors that winemakers need to monitor, from characteristics of the grapes all the way to the nature of the wine once the bottle is opened a considerable period of time later.

An application close to final consumption is barrel tasting in which a wine is previewed well in advance of bottling while it is still ageing in barrel (Figure 6.1). Glass pipettes are routinely used to extract the wine and then deposit in a glass for tasting. A reasonably experienced palate is required to analyze the wine in this embryonic state; one needs to know how a given character in barrel will evolve a year or more later in bottle. The eminent wine importer Kermit Lynch has publicly lamented the increasing use of barrel tastings as a manifestation of newer wine connoisseurs' impatience with the by-product of producing "New World" style wines that are drinkable upon first opening. Lynch feels that the "Old World" style

The Glass of Wine: The Science, Technology, and Art of Glassware for Transporting and Enjoying Wine, First Edition. James F. Shackelford and Penelope L. Shackelford.

Figure 6.1 Giuseppe Parlavecchio, viticulturist at Pietradolce winery near Mt. Etna in Sicily, provides a barrel tasting of their Etna Rosso. (Reproduced with kind permission of Giuseppe Parlavecchio, Pietradolce.)

(at least for reds such as a traditional Bordeaux) involved a less refined vinification that required substantial aging and subsequent aeration upon uncorking. Of course, glass is the innocent bystander in this debate, simply providing its pristine surface to those wishing to preview a wine that will be available a year or two down the road.

Glass' friendly surface has led to other sundry roles around the winery. Figure 6.2 shows an "airglass bubbler" on top of an oak barrel at the Fontodi Winery in the Chianti region of Northern Italy, a winery that has been especially renowned for its sangiovese-based wines, including Chianti. Ironically, Fontodi has been owned since 1968 by the Manetti family that has for centuries been associated with the manufacture of the famous ceramic tile from the Chianti region, a product closely related in nature with the ceramic amphorae so often mentioned in this book. In the airglass bubbler application, glass provides a visual confirmation that the barrel (or amphora) is adequately filled with wine, as well as a monitor of variations in the wine level over time.

Beyond the miscellaneous glassware around the barrel rooms, a concentration of **scientific glassware** is invariably found in the winery's laboratory (Figure 6.3). All but the smallest operations are required to do substantial chemical analyses during the winemaking process.

A critical test required during harvest is the **°Brix** (pronounced "bricks") **number** measurement that indicates the sugar content of the grape juice (which of course becomes the alcohol upon fermentation). As a practical matter, the quantity that is measured is the density of the grape juice that indicates the amount of soluble solids per unit volume of juice. This simple indication assumes the juice is a mixture of solids and water, with sugar being 90–95% of

Figure 6.2 An airglass bubbler on top of an oak barrel at the Fontodi Winery in the Chianti region of Northern Italy provides confirmation that the barrel is adequately filled with wine. (Reproduced with kind permission of Azienda Agricola Fontodi.)

Figure 6.3 Even at a relatively small boutique winery, such as Quinta Vale D. Maria in the Duoro Valley of Portugal, a laboratory with a variety of scientific glassware is essential. In this case, the laboratory also serves as a tasting room, with an impressive selection of stemware. (Reproduced with kind permission of Cristiano van Zeller, Quinta Vale D. Maria.)

the total soluble solids. The traditional device for this measurement is the *hydrometer*, as illustrated in Figure 6.4. This simple piece of **laboratory glassware** uses the classic Archimedes' principle of buoyancy to determine the density of the juice. In essence, the device indicates that the juice has a density greater than that of pure water. The density increase can then be calibrated against the sugar content (°Brix in units of grams of sugar/100 g of juice). For convenience, the hydrometer can be scaled in the °Brix units rather than density

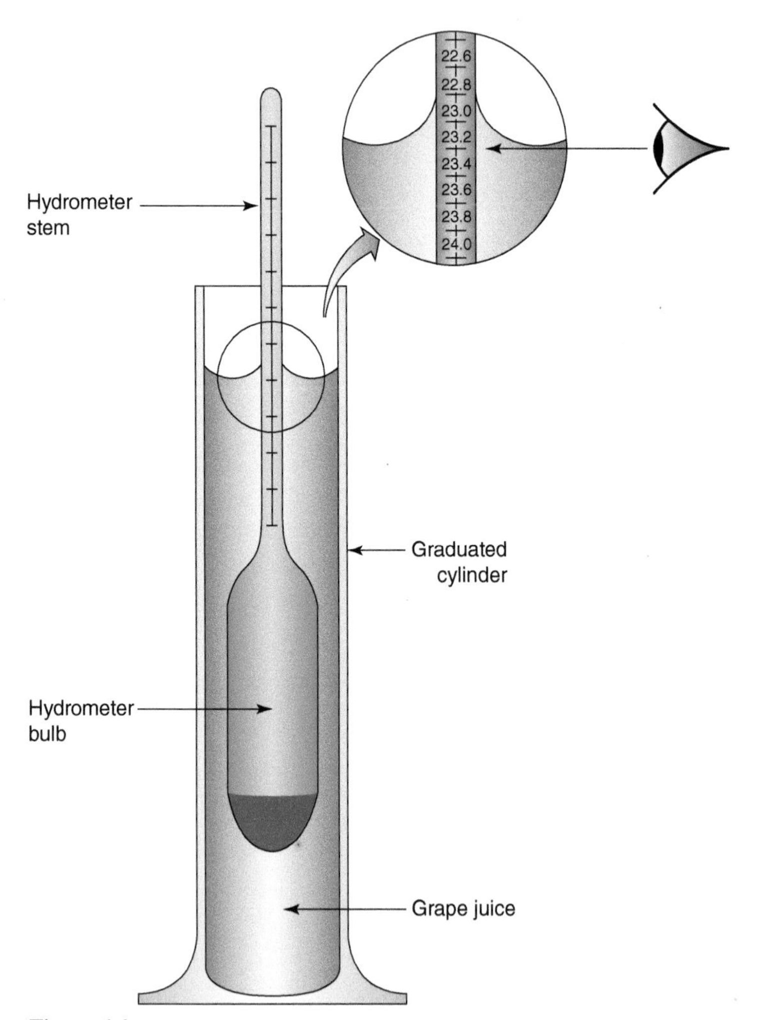

Figure 6.4 A glass hydrometer is the traditional instrument used to measure the °Brix number (sugar content) for grape juice. The hydrometer is fundamentally measuring the density of the juice, a sensitive function of sugar content. This information is critical to indicate the optimal time of harvest and to estimate the resulting alcohol content of the final wine. In this illustration, the hydrometer stem is calibrated in the °Brix number in units of grams of sugar/100 g of juice.

allowing the sugar content to be read directly. Figure 6.4 shows such a reading for a °Brix of 23.3 for juice at a level typical of grapes approaching harvest time.

Not only does the presence of dissolved sugar produce a measureable increase in density of the juice (compared to pure water), the sugar also changes the optical behavior of the liquid in a way measureable by a *refractometer*. This complementary technique specifically measures the bending of light (*refraction*) passing through the juice, a sensitive function of the sugar content. We will return to the topic of refraction in detail in Chapter 10 as it is one of the central characteristics of the optical behavior of glass. The principle of the refractometer is illustrated in Figure 6.5, with Figure 6.5a showing how, for an arbitrary incident angle of 45°, the degree of bending of light as it passes from air into liquid varies slightly with the sugar content of the liquid and Figure 6.5b shows a simplified schematic of a typical hand-held refractometer in which the bending of light through a glass prism and lens produces an easy to read output of the °Brix. (Actual devices frequently employ a more elaborate optical system of multiple prisms and lenses.)

Laboratory glassware is also used to monitor the acidity of the wine through pH, titratable acidity (TA), and volatile acidity (VA) measurements. While pH is an overall measure of acidity and an indicator of the wine's stability and quality, the TA focuses on specific organic acids, primarily tartaric acid – a strong indicator of the wine's taste. In addition, the partially soluble salts of tartaric acid, potassium bitartrate and calcium tartrate, contribute to the wine's stability. As a result, a thorough evaluation of the acidity of both the grape juice and the final

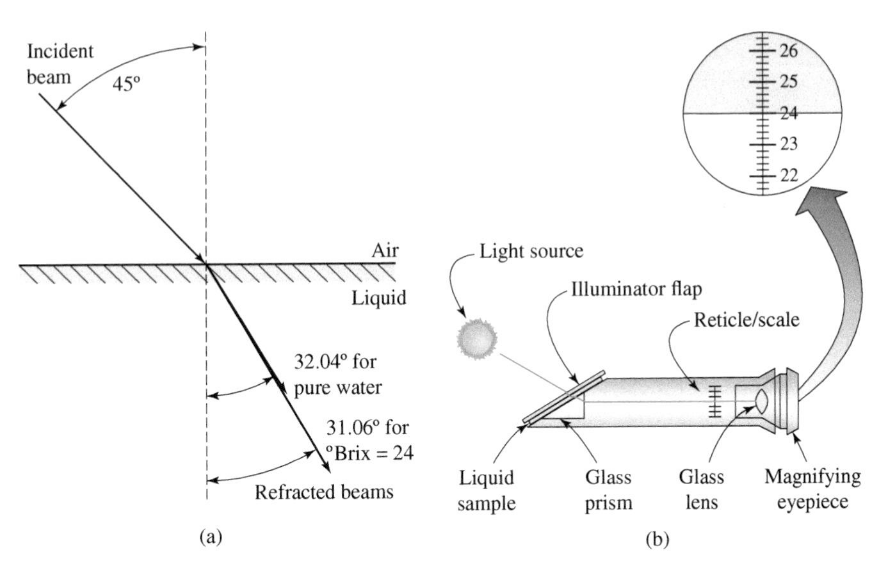

Figure 6.5 (a) The bending of light (refraction) as it passes from air into water increases slightly with increasing sugar content of the liquid. (b) A schematic of a hand-held refractometer in which the refraction of light through a glass prism and lens produces a readout of the °Brix through the eyepiece. The optics of the device allow the slight change in refraction to display °Brix within ±0.1.

wine is important. As important as this information is, it is also limited in nature. Both pH and TA measurements are overall measurements, and nonunique combinations of different acids can give identical numbers. Nonetheless, the information as general as it might be is essential to provide guidance to the winemaker.

As you might recall from an introductory chemistry class, pH represents the hydrogen ion (H^+) concentration in an aqueous solution, with pure water having a neutral value of 7 and "acidic" solutions having values less than 7 and "basic" solutions having values greater than 7. The lower the number, the more acidic the solution; the higher the number, the more basic. The pH meter, now a staple of contemporary chemistry labs in the wine industry and beyond, has in fact an interesting history, based on glass. The concept of using glass electrodes for pH measurement was developed by Fritz Haber (an early winner of the Nobel Prize in Chemistry for his pioneering work on nitrogen fixation) and one of his students in 1909, but it was the American chemist Arnold Beckman who turned the concept into an electronic instrument with his 1936 patent that connected the glass electrodes to a vacuum tube amplifier and voltmeter that provided a sensitive measurement of the pH (Figure 6.6). The voltage generated between the two electrodes is directly proportional to the pH. Beckman's invention (originally

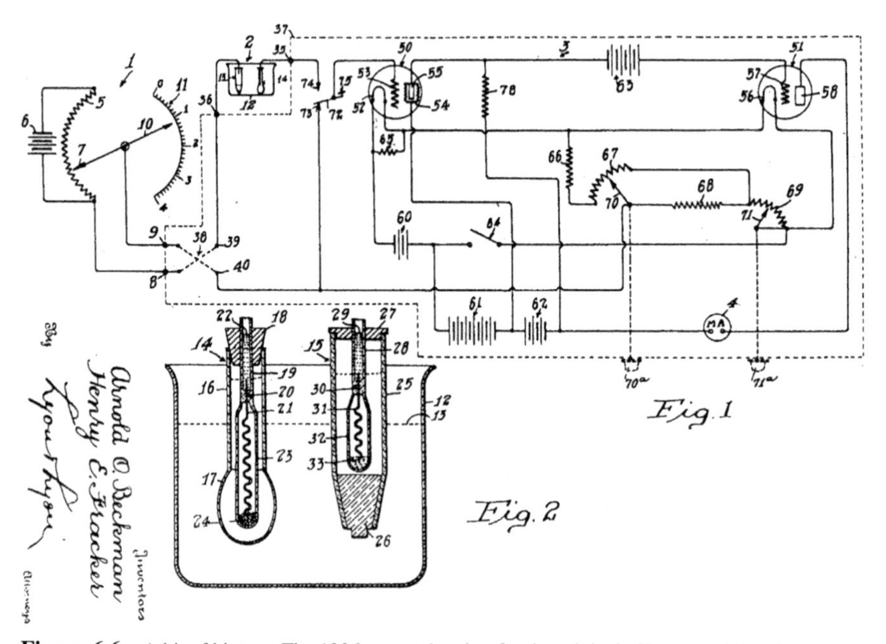

Figure 6.6 A bit of history: The 1936 patent drawing for the original pH meter of Arnold Beckman. The heart of the system is the two glass electrodes (14 and 15) located in the lower part of the drawing (identified as "Fig. 2" in the original). The external circuit measures the voltage generated between the electrodes when immersed in a solution, and the voltage is directly proportional to the solution's pH. (US Patent 2,058,761, October 27, 1936).

called the acidimeter) led to the development of Beckman Instruments with its subsequent role as a major supplier of scientific instrumentation throughout the twentieth century. Dr. Beckman's legacy lives on in both the Beckman Center, the Western US home of the National Academies, and the Beckman Laser Institute, both located at the University of California, Irvine. Beckman's design, originally developed to measure the acidity of citrus crops in Southern California, is the basis of modern pH meters that are now more compact and digitized. For wine grape juice, early maturity fruit can have pH values of 2.8–3.0. A desirable range for table wines is from 3.0–3.3.

The second important acid measurement is titratable acidity, which requires us to describe *titration* for those who are not familiar with this analytical chemistry technique. Titration originated in the eighteenth century and has become a widely used and routine experimental tool in contemporary chemistry labs. In brief, a reagent is systematically added to a precisely measured volume of the solution being analyzed until a standard endpoint is reached. The volume of reagent required to reach the endpoint indicates the concentration of the solution being analyzed. For the case of grape juice or wine, the acidic sample is titrated to an endpoint of $pH = 8.2$ (in the United States, with different endpoints used in some other wine producing countries) using a strong base as the reagent. Titration is, in fact, used in a variety of experiments around the wine laboratory, as will be shown later in relation to the measurement of sulfite levels.

Volatile acidity is primarily associated with acetic acid ($C_2H_4O_2$), the main volatile acid in fermented beverages, and acetic acid is associated with wine spoilage. VA analysis generally involves steam distillation, and the collected volatile acid distillates are titrated with sodium hydroxide, NaOH. The results are given in grams/liter acetic acid. A glassware system for this purpose is shown in Figure 6.7.

Nearly every bottle of wine sold in the United States has the phrase "contains sulfites" on its label. While many consumers may ignore the phrase, others are curious and some are concerned, occasionally highly so. Rumors of sulfites causing headaches and/or rashes are widespread. Karen MacNeil in her remarkably thorough coverage of wine in the appropriately titled *Wine Bible*, attempts to dispel these concerns and traces the anxiety over sulfite levels back to the rise in popularity of salad bars in the 1970s, where large amounts of sulfites were sprayed on freshly cut fruits and vegetables to delay spoilage. Unfortunately, a very small fraction of the population (primarily asthmatics) had adverse reactions to the presence of these sulfites. Federal regulations followed and, in 1988, were extended to wine with a limit of 350 parts per million (ppm). The cautionary label "contains sulfites" is required for any wine having more than 10 ppm sulfites, a threshold crossed by nearly all wines, given that sulfites are a natural by-product of the fermentation process. Nonetheless, sulfur dioxide (SO_2) is often added to wine for its antimicrobial benefits that help prevent spoilage, as well as oxidation. Even with these controlled additions, the level of sulfites in most wines falls below 100 ppm.

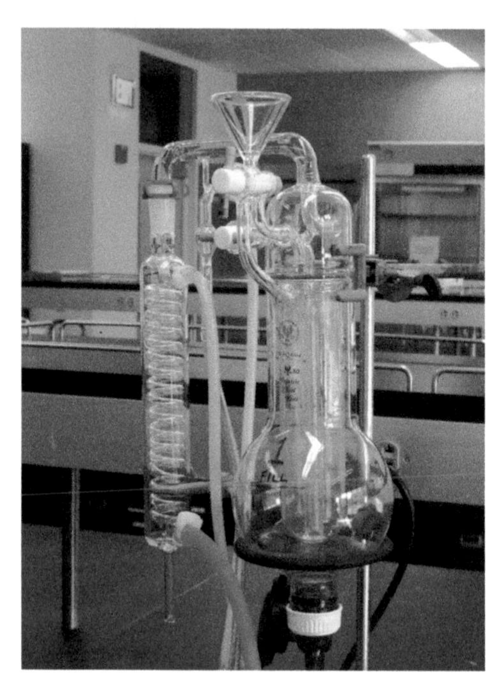

Figure 6.7 The configuration of glassware for determining the volatile acidity (VA) of wine, an indicator of the amount of acetic acid and corresponding spoilage. (Reproduced with kind permission of Andrew L. Waterhouse, Robert Mondavi Institute for Wine and Food Science.)

As a practical matter, the "sulfite" level is measured as the total sulfur dioxide content by any of a variety of methods. (Sulfite and other ionized species are produced upon the dissolution of SO_2 in the juice or wine.) A glassware system for this purpose, known as the aeration/oxidation (AO) procedure, is shown in Figure 6.8. This system involves the distillation of the SO_2 out of the sample into a hydrogen peroxide (H_2O_2) trap. There, the SO_2 is oxidized to sulfuric acid H_2SO_4 that is titrated against the base sodium hydroxide (Figure 6.9).

An especially important test whose result is of interest to the winemaker as well as the consumer is the alcohol content of the final wine, given in volume %, a datum also found on any wine label. In a sense, this is the "bottom line" for many consumers and at the heart of a continuing debate among oenophiles – how much alcohol is too much? For some guidance, we can return to the comments of wine importer and merchant Kermit Lynch at the opening of this chapter. As we noted, Lynch is a great advocate of the "Old World" style of wine, meaning that typified by the traditional wines from Europe with distinctive acidity and relatively low alcohol levels around 13 volume % or less. On the other hand, "New World" wines are those with characteristically more bold flavors and often high alcohol levels. Some California cabernet sauvignons can be found with alcohol levels around 15%, and California zinfandels even higher!

Figure 6.8 This glassware system is used for the aeration/oxidation (AO) process for the measurement of total SO_2 content of wine, an indication of the sulfite content. (Reproduced with kind permission of Andrew L. Waterhouse, Robert Mondavi Institute for Wine and Food Science.)

Many have blamed the prominent and highly influential wine critic Robert Parker for this phenomenon and refer with some derision to "Parkerized" wines. A New York counterpart to the Berkeley-based Kermit Lynch is the wine merchant Neal Rosenthal who imports a wide range of European wines, often from small, family producers. In the 2004 documentary on the effect of globalization on the wine industry entitled *Mondovino*, Rosenthal speaks of Parker in satanic terms for his singular influence in moving the industry toward "New World" values. Beyond this ongoing and often heated debate over the style of wine, we must also note an important pragmatic consideration, namely, the level of taxation on wine is primarily based on the alcohol content.

Again, glass is an innocent bystander in both this debate and the eventual tax burden tied to the level of alcohol. A traditional technique for determining alcohol level is the ebulliometric analysis, a direct measurement of the boiling point of the liquid that takes advantage of the well-defined relationship between the amount of boiling point suppression and alcohol content. A table wine with a relatively high alcohol content of about 15 volume % can reduce the boiling point of an aqueous solution by 10 °C (down from the boiling point of pure water at 100 °C). Other than the use of a glass thermometer, the ebulliometer is largely a metallic device. A more contemporary method for measuring alcohol

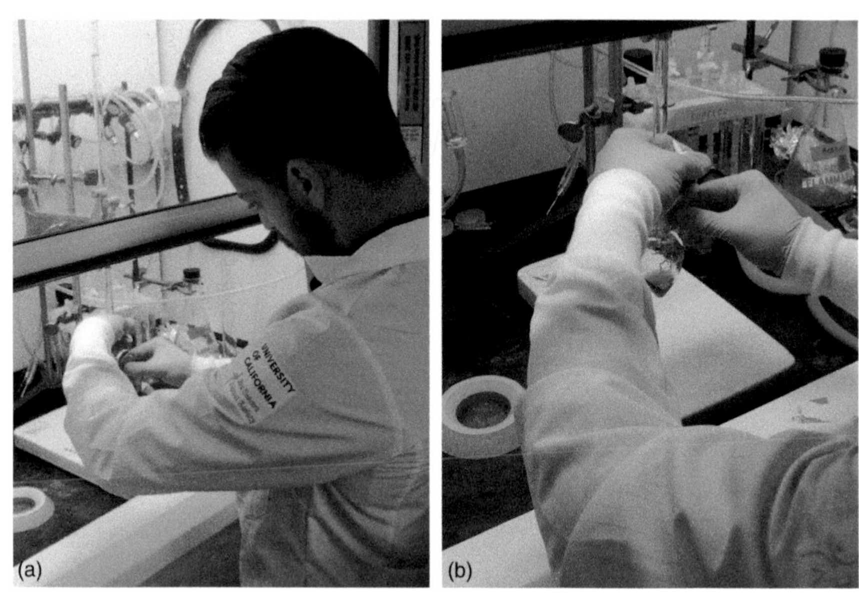

Figure 6.9 (a) Luigi Picariello of the Università Degli Studi di Napoli Federico II is shown completing the aeration/oxidation (AO) process for the measurement of total SO_2 content of wine by (b) titrating the H_2SO_4 produced in the process against the base sodium hydroxide. (Reproduced with kind permission of Andrew L. Waterhouse, Robert Mondavi Institute for Wine and Food Science.)

content is to use a gas chromatograph (GC) to measure ethanol. The GC is a widely used analytical chemistry system that separates and analyzes compounds (such as ethanol) that have been volatilized from a sample (such as wine). As with the ebulliometer, the GC is not primarily manufactured from glass. Nonetheless, the glass insert shown in Figure 6.10 is an essential component for

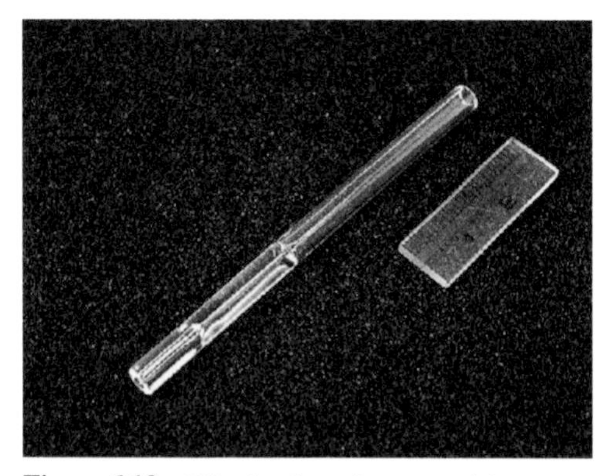

Figure 6.10 This glass insert is an essential component in the gas chromatograph used to measure the alcohol content of wine samples. (Reproduced with kind permission of Kevin McLaughlin, Shimadzu.)

preventing the contamination of the GC column by sample components. The regular exchange of the insert is essential to avoid poor reproducibility of results due to the crossover of residual samples.

You may have noticed the use of the terms "laboratory glassware" and "scientific glassware" in this chapter. Is this material used to analyze juice and wine different than the glass used for bottling and drinking wine? In a word, *yes!* The glass developed for chemical laboratories is a **borosilicate** in which a significant amount of B_2O_3 is part of the glassmaking batch. The B_2O_3 is, similar to SiO_2, a glass former contributing to the chemical durability of the glassware. At the same time, about 4 wt% Na_2O provides good formability at economical manufacturing temperatures without sacrificing the durability provided by the glass forming oxides. Nonetheless, this Na_2O content is considerably lower than the 13 wt% in typical bottles. This lower Na_2O level contributes to the greater chemical durability of the glass. Equally beneficial is a high SiO_2 content of 80 wt% compared to 72 wt% in a typical glass bottle. In addition, there is another 13 wt% B_2O_3, bringing the total content of glass formers to a robust 93 wt%. There are many commercial producers of this type of laboratory glassware with Corning's Pyrex® being the best known. Another by-product of this particular glass composition is a low value of thermal expansion that contributes to the material's ability to survive rapid temperature changes or **thermal shock**, usually the result of a rapid cooling. The concept of thermal shock will be discussed further in Chapter 13 in relation to a creative way of opening wine bottles. The resistance to thermal shock also makes Pyrex® and similar brands popular for use in the home kitchen in which casserole dishes routinely survive on being removed from a hot oven.

So, we see that glass plays myriad supporting roles around the winery, from the glass in the pipettes used for barrel tastings to high-quality optical prisms in the scientific instrumentation that functions alongside numerous pieces of laboratory glassware. Next, we delve deeper in our understanding of glass, surveying the sophisticated technology involved in modern glassmaking.

BIBLIOGRAPHY

Boulton, Roger, Vernon Singleton, Linda Bisson, and Ralph Kunkee, *Principles and Practices of Winemaking*, Springer, New York (1999).

Iland, Patrick, Nick Bruer, Greg Edwards, Sue Weeks, and Eric Wilkes, *Chemical Analysis of Grapes and Wine: Techniques and Concepts*, Patrick Iland Wine Promotions, Campbelltown, Australia (2004).

MacNeil, Karen, *The Wine Bible*, 2nd Edition, Workman, New York (2015).

McCoy, Elin, *The Emperor of Wine: The Rise of Robert M. Parker, Jr. and the Reign of American Taste*, HarperCollins, New York (2005).

Rosenthal, Neal, *Reflections of a Wine Merchant*, Farrar, Straus, and Giroux, New York (2008).

Shackelford, James, *Introduction to Materials Science for Engineers*, 8th Edition, Pearson, Upper Saddle River, NJ (2015).
Smrček, Antonín, "Compositions of Industrial Glasses," in *Fiberglass and Glass Technology*, Frederick Wallenberger and Paul Bingham, Eds., Springer, New York (2010).
Zoecklin, Bruce, Kenneth Fugelsang, Barry Gump, and Fred Nury, *Wine Analysis and Production*, Chapman and Hall, New York (1995).

Chapter 7

Perfection Through Fire – Modern Glassmaking

"Who, when he saw the first sand or ashes, by a casual intenseness of heat, melted into a metalline form, rugged with excrescences, and clouded with impurities, would have imagined, that in this shapeless lump lay concealed so many conveniences of life, as would in time constitute a great part of the happiness of the world?"

Samuel Johnson in *Rambler #9* (April 17, 1750)

In Chapter 3, we marched quickly through four millennia of glassmaking, arriving at the current multibillion dollar industry that includes a vast number of wine bottle and stemware manufacturers. As noted there, much of the industry involves converting locally available deposits of sand (SiO_2), soda ash (Na_2CO_3), and lime (CaO) into these durable and transparent vessels (Figure 7.1). The energy-consuming fire of the modern furnace converts the soda ash into soda (Na_2O) and gaseous carbon dioxide (CO_2) with the bubbles of CO_2 gas mixing together the remaining ingredients into the homogeneous "soda-lime-silica glass," while, in the process, providing an abundance of climate-threatening carbon dioxide. These furnaces can be in the service of individual artisans who make bottles and stemware by hand and blowpipe or under the control of plant engineers commanding massive automated machinery.

Descendents of Michael Owens' pioneering machine design that we introduced in Chapter 3 now produce most of the contemporary bottles (Figure 7.2). The techniques used to rapidly form bottles have also been adapted to producing stemware at substantially lower prices compared to hand-blown pieces but with impressively high-quality craftsmanship (Figure 7.3).

The Glass of Wine: The Science, Technology, and Art of Glassware for Transporting and Enjoying Wine, First Edition. James F. Shackelford and Penelope L. Shackelford.

Figure 7.1 The glassware that stands ready for an evening of fine dining is the result of a technology no less complex than that required to produce the wine itself. This substantial collection of stemware and other glass vessels is at the Quince Restaurant in San Francisco, CA. (Reproduced with kind permission of Zoe Simonneaux, Quince Restaurant.)

Figure 7.2 The automated production of bottles was a triumph of both materials science and mechanical engineering in the twentieth century, as illustrated in Figure 3.11. The result of this technology is that now contemporary wineries have an abundance of economical and uniform bottles ready for filling, in this case at the Bele Casel Winery in the Prosecco region of Northern Italy. (Reproduced with kind permission of Società Agricola Bele Casel.)

Figure 7.3 An elegant piece of stemware is formed, not by the skill of a glassblower but by the precision of mechanisms within the heart of a massive machine at the Luigi Bormioli factory in Parma, Italy. (Reproduced with kind permission of Kim Goodwin, Luigi Bormioli, Italy.)

Whether glass bottles or stemware are made by hand or by machine, whether one at a time or by the thousands, the glassmaker's art requires a delicate control of the **viscosity** of the glass material itself. For our purposes, we can define viscosity as simply the resistance to deformation. Most have observed a glassblower at a crafts fair making small figurines. By applying heat from a torch to small rods of glass, the artisan can manipulate the glass material very differently at different temperatures. Under the direct heat of the torch, the yellow-hot material bends and slumps quickly to the artist's will. Remove the torch, and the glass cools and again becomes transparent and the shaping process continues as the material stiffens as the cooling quickly proceeds. Less obvious perhaps, the glassmaker puts the finished object (a small swan, for example) into a warm oven and lets it sit there for several minutes before putting it out for sale. This final **annealing** stage is required to relieve stresses created in the rapid heating and cooling operations of melting and forming. All of these steps can be mapped along a "viscosity curve" that quantitatively summarizes the level of viscosity at the various temperatures involved in the overall process (Figure 7.4).

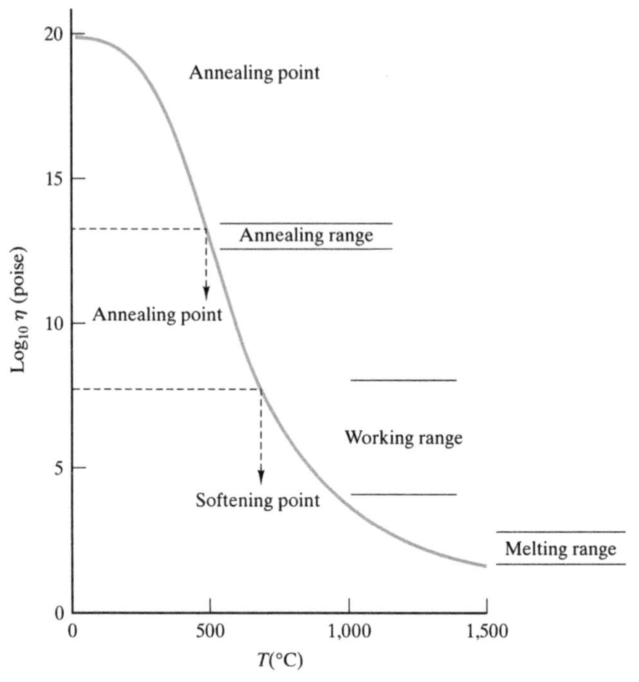

Figure 7.4 The wide range of viscosities over which glass behaves from room temperature to its melting range guides the glassmaker in forming bottles and stemware. (Shackelford, 2015. Reproduced with permission of Pearson.)

The physical unit of viscosity is the "poise" named in honor of the nineteenth century French physician and physiologist Jean Louis Marie Poiseuille who did pioneering work on blood flow. The range of viscosity is enormous, and the values in poise are plotted on a logarithmic scale (by powers of 10) to reflect this fact. At room temperature where the glass is a rigid solid, the value of viscosity is very large, about 10^{20} poise. When the glass is raised to its **melting range** (to produce, for example, the molten gob of glass that will become a bottle), the viscosity drops dramatically to about 10^{2} poise. This relatively thick liquid has a consistency comparable to honey. By contrast, the viscosity of water is even smaller, about one hundredth (10^{-2}) of one poise. The more viscous (thicker) honey-like gob is convenient for manipulation in a bottle mold and subsequent blowing and forming. The glass cools rapidly as it falls into the mold and the air jet forms the first rough bottle shape. This **working range** corresponds to intermediate temperatures between 700 and 900 °C and viscosities between 10^{4} and 10^{8} poise. As with the crafts fair artisan, the manufacturer of wine bottles must anneal the finished products at a relatively low temperature so that the internal stresses created by the rapid heating and cooling steps in forming processes can be relieved over a period of about 15 min. The **annealing**

Figure 7.5 The artisan producing this elegant example of modern stemware is, in effect, performing a carefully choreographed dance along the viscosity curve of the glass material, as it rapidly heats and cools over the melting, working, and annealing ranges. (Reproduced with kind permission of Riedel Glas, Austria.)

range corresponds to a temperature around 450 °C and a viscosity of about 10^{13} poise. This elaborate journey along the viscosity curve of Figure 7.4, as controlled by a sophisticated bottle-making machine, is essentially the same in the hands of a skilled artisan producing elegant stemware (Figure 7.5).

While we concentrate on soda and lime as the primary components added to sand in making common glasses, some other oxides are widely used for useful characteristics they provide. Lead oxide is an especially popular example for quality stemware known as "lead crystal." We recall our gratitude to George Ravenscroft in Chapter 3 for his seventeenth century invention, as well as the unfortunate complication of nomenclature between "crystal" in this sense and the fact that, on the atomic structural scale, glass is *noncrystalline* with its atoms arranged in a largely random fashion (Figure 3.5). As noted in Chapter 3, the "crystal" nature of lead crystal is its high degree of sparkle when viewed under ordinary light, an optical effect that we often associate with fine gems that are crystalline minerals with high *indices of refraction*. (See Chapter 10 for a discussion of these indices and how they correspond to the ability to bend light and produce this sparkle.) So, this beautiful *noncrystalline* stemware sparkles like a crystalline gemstone.

Of course, having 24 wt% PbO by weight in the typical chemistry of fine stemware can provide some pause for thought. Lead (Pb) by itself is a notoriously toxic element. Many decades ago, it was not uncommon to obtain cheap pottery with shiny glazes produced by the generous use of PbO in the glaze composition. The use of this pottery in serving food and beverages too often led to the leaching out of some of the lead with considerable health consequences. Regulatory agencies have largely eliminated this problem with careful testing and inspection programs. Fortunately, the lead contained in the lead oxide of glass stemware is strongly bound and has been shown not to be a health hazard. Only in the exceptional case of wine stored in a lead glass decanter for more than a month might one find troubling levels of lead leached into the beverage. In reality, the stale wine would be undrinkable in most cases, not because of the lead content but because of the ravages of oxidation. An exception can be the temptation to store Cognac or Port in a lead crystal decanter. Such wines with substantial amounts of residual sugar could withstand a long storage time, and the chance of some lead leaching would be a legitimate concern. In summary, lead crystal stemware is not a safety concern, but lead crystal decanters should be avoided for long-term wine storage.

Decorative colors and adornments on stemware have largely fallen out of fashion over the past several decades (Figure 7.6). The optically transparent, simple tapered bowl styles popularized by Riedel designs in the 1960s are now

Figure 7.6 These elegant Hawkes wine glasses produced for White House collections from the 1920s to the 1950s violate most of the design features of modern stemware. The coloring, cuts, and other patterns serve to obscure our view of the wine. The vertical walls or outward flare of the bowls fails to capture the aroma of the wine, as do modern tapered bowls. (Reproduced with kind permission of the Corning Museum of Glass.)

commonplace both at home and in restaurants. These uncolored walls and simple shapes allow connoisseurs to evaluate the wine with minimal distractions.

On the other hand, the coloration of bottles continues long-standing traditions. Even many white wines are stored in colored bottles, often deeply colored. Glass is generally colored by the addition of small amounts of so-called **transition metal oxides**. Among the most common examples is iron oxide for brown and green colors (the common beer bottle is an example, along with bottles for many wine varietals). We will return to these common examples in Chapter 10 as part of our discussion of the optical properties of glass, but first we will illustrate coloring with cobalt oxide that produces a distinctive blue color well known as "cobalt blue" and a color widely used in decorative ceramic glazes. While blue wine bottles are relatively uncommon, they are used and are, in fact, rather popular for certain rieslings from Germany and elsewhere, as shown in Figure 7.7.

The way transition metal oxides work is a simple lesson in optics. The human eye can perceive wavelengths in the electromagnetic spectrum between about 400 and 700 nm. When the eye perceives a spray of photons in roughly equal numbers across this entire "visible range," we perceive that as "white light." On the other hand, individual wavelengths across the spectrum literally

Figure 7.7 While blue wine bottles are relatively uncommon, they are used for a number of German rieslings that occasionally even have "blue" as part of their identity.

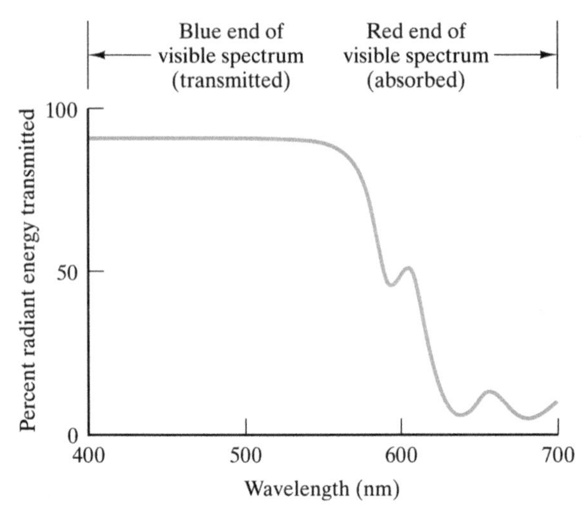

Figure 7.8 Only 1% cobalt oxide (CoO) added to an otherwise transparent silicate glass leads to the red end of the visible light spectrum being absorbed leaving a characteristically blue light being transmitted. (Shackelford, 2015. Reproduced with permission of Pearson.)

provide the colors of the rainbow, from blue on the low (400 nm) end of the spectrum to red on the high (700 nm) end. By adding a small amount of cobalt oxide (CoO) to an otherwise transparent glass composition, the Co^{2+} ions absorb wavelengths in the red end of the spectrum leaving a distinctive blue color transmitted to the eye of the observer (Figure 7.8). The mechanism by which the Co^{2+} ions absorb red light is electron energy transitions, an effect into which we will dwell more deeply in Chapter 10. In this regard, the term "transition" also comes from the positions of these elements in the periodic table and the arrangement of their electrons in their planetary orbits around their atomic nuclei, topics that will be reviewed in Appendix A.

The exact nature of the color produced by a particular transition metal oxide depends on the exact nature of the atomic-scale environment in which a particular **transition metal ion** finds itself, specifically the number and arrangement of oxygen ions in its immediate vicinity. Cobalt ions can produce coloring from the characteristic "cobalt blue" to a pinkish character. Iron ions can produce colors from deep brown to blue-green to a weak yellow, again depending on the exact arrangement of oxygen ions surrounding the iron atom. Again, the nature of color formation will be explored in greater depth in Chapter 10.

An increasingly popular method of labeling wine is to forgo the traditional paper label and use colored glass designs. Elaborate multicolored designs in glass-containing inks are produced on paper-backed decals that are then applied to the bottle surface and subsequently fired at relatively low temperatures. A typical firing temperature for these decals is 650 °C. Polymeric matrices for the

Figure 7.9 This wine bottle label is a detailed pattern of colored glass produced by firing a decal onto the bottle at a temperature below that originally used to form the bottle itself. (Reproduced with permission of Tina Ficarra, Bergin Screen Printing and Etching.)

colored glass particles burn away during the firing leaving behind durable glass-on-glass designs (Figure 7.9). We see from the viscosity curve that the decal firing temperature is slightly above the annealing temperature but comfortably below the working range to prevent any distortion of the bottle shape in this final labeling step.

Finally, as we noted in Chapter 3, glass manufacturing is a major energy-consuming industry, but the level of consumption can be tempered by the **recycling** of used glass. Recycling also reduces the consumption of raw materials. In general, the use of recycled glass contributes to increased production rates and reduced pollution emissions. Broken chunks of used glass are called **cullet** and can constitute up to 95% of the glass batch in new bottle manufacturing. The actual amount depends on many factors including whether variations in the cullet might undermine the final product quality, especially color and transparency.

And so now, we hope the making of glass is clear. We see more fully how modern glass bottles and stemware are made to the highest standards by hand as well as by machine. But, as we look at these vessels, new questions arise. Why is that bottle of burgundy so much heavier than a sauvignon? Will a cabernet drunk from a chardonnay glass not be the same as the one specifically designed for Bordeaux-like reds? We will explore the shape of bottles and stemware (stem-less ware too!) in Chapters 11 and 12, but, in the meantime, the following three chapters take us further into the field of glass science to explore in greater depth the atomic- and microscopic-scale structure of the material, along with the resulting mechanical and optical properties.

BIBLIOGRAPHY

Kingery, W. David, H. Kent Bowen, and Donald Uhlmann, *Introduction to Ceramics*, 2nd Edition, John Wiley & Sons, Inc., New York (1976).

Shackelford, James, *Introduction to Materials Science for Engineers*, 8th Edition, Pearson, Upper Saddle River, NJ (2015).

Smrček, Antonín, "Compositions of Industrial Glasses," in *Fiberglass and Glass Technology*, Frederick Wallenberger and Paul Bingham, Eds., Springer, New York (2010).

Vogal, Werner, Norbert Kreidl (Translation Editor), and Ester Lense (Associate Editor), *Chemistry of Glass*, American Ceramic Society, Westerville, OH (1985).

Chapter 8

Beauty of a Random Nature – Glass Structure on the Atomic Scale

"It must be frankly admitted that we know practically nothing about the atomic arrangement in glasses."

W.H. Zachariasen – as stated in the opening sentence of his seminal paper on the subject of atomic arrangement in glasses in the *Journal of the American Chemical Society,* **54**, 3841 (1932).

Prior to William Houlder Zachariasen's publication in 1932, his opening sentence was by and large correct. That particular paper, on the other hand, gave us an understanding of the nature of glass structure on the atomic scale that has served us well to date. Zachariasen was a surprising source of the most important paper in the history of glass science. The Norwegian-American physicist spent his career primarily working on the use of X-rays to study the structure of a wide range of crystals. We need to recall the strong distinction between the regular order of crystals and the random nature of noncrystalline glasses introduced in Figure 3.5 to fully appreciate how, in a sense, Zachariasen was a fish out of water in taking on the question of how individual atoms are arranged in a glass. Nonetheless, his 1932 paper argued correctly that, as Figure 3.5 illustrated in a two-dimensional schematic way, the basic "building block" of silicate minerals is a tetrahedral structure composed of four oxygen atoms surrounding a central silicon atom (Figure 8.1) and that building block is preserved in a noncrystalline glass of the same chemical composition. The

The Glass of Wine: The Science, Technology, and Art of Glassware for Transporting and Enjoying Wine, First Edition. James F. Shackelford and Penelope L. Shackelford.

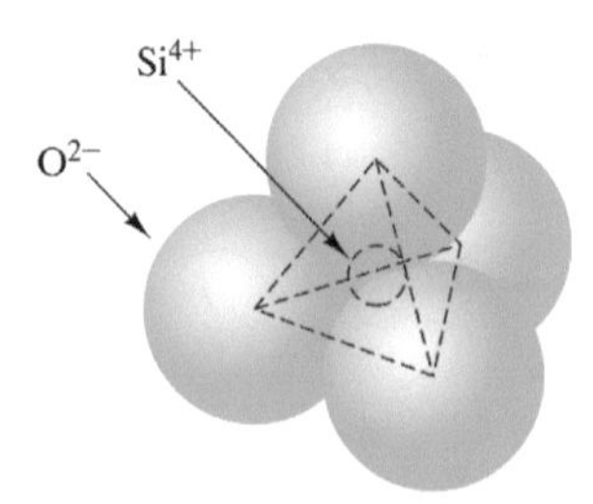

Figure 8.1 The basic "building block" of silicate minerals is a tetrahedral cluster of four oxygen atoms surrounding a central silicon atom. Zachariasen argued correctly that the same local structure would be true in silicate glasses, although the linkage of these building blocks would be random unlike the regular order in a crystalline mineral. (Shackelford, 2015. Reproduced with permission of Pearson.)

difference between the crystal and the glass is then the nature of the linkage of the adjacent silica tetrahedra. In the crystal, the linkage is regular and repeating. Figure 3.6 showed such a pattern for common quartz, the room-temperature crystallographic form of silica (and representing the atomic-scale structure of common sand), and Figure 3.7 showed a computer-generated linkage of such tetrahedra for pure silica glass. Figure 8.2 shows another, similar comparison. In this case, **cristobalite**, the high-temperature crystallographic form of silica (stable just below the **melting point**), is compared with the random network of

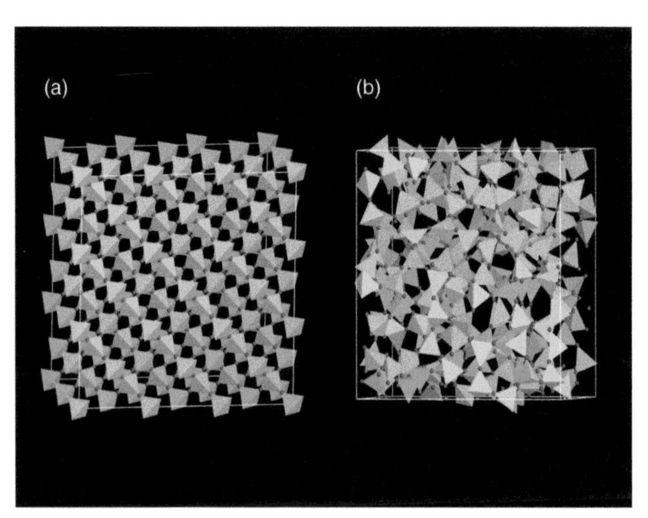

Figure 8.2 A comparison of computer-generated models of (a) the high-temperature crystal form of silica, cristobalite, and (b) silica glass (an image equivalent to Figure 3.7). Together, these two images serve as a three-dimensional version of the two-dimensional schematic in Figure 3.5 comparing the structures of a crystal and its chemically equivalent glass side by side. (Reproduced with kind permission of Sabyasachi Sen, Department of Materials Science and Engineering, University of California, Davis, CA.)

silica glass. Some glass scientists argue that the structure of cristobalite is closer in nature to that of silica glass than is quartz because of the fact that it is the crystallographic form stable just before the material melts beginning the process of forming a glass, as well as the fact that cristobalite and silica glass have similar densities. Figure 8.2 is also a three-dimensional version of the two-dimensional schematic in Figure 3.5, and, we would argue, *a beautiful one!*

Before going on with our discussion of atomic structure, we need to comment on what we (and before us Zachariasen) mean by "atom." While we shall in general follow Zachariasen's lead and refer strictly to atomic structure throughout this chapter, many readers will be aware that the atomic bonding in ceramics and glasses can be described as ionic in nature, and so the "atoms" are in fact more precisely described as **ions**. For those who wish an introduction (or refresher) on the various forms of atomic bonding, that information is provided in Appendix A.

The Zachariasen description of glass structure illustrated in a contemporary way in Figure 8.2b is also known as the **random network model**. Zachariasen's depth of knowledge about the atomic arrangement in silicate minerals allowed him to go beyond the important statement that crystals and glasses would tend to have the same building block structure. He also described the way in which those building blocks are linked. As we see by looking closely at both crystalline silica and noncrystalline silica (Figure 8.2b for silica glass), the oxygen atoms at the corners of each tetrahedron are the "bridges" that connect the adjacent tetrahedral clusters (Figure 8.3).

Since the advent of substantial and relatively economical computer power in the 1970s, a new branch of the field of materials science has evolved: *computational materials science*, and the steady growth of computation capacity at

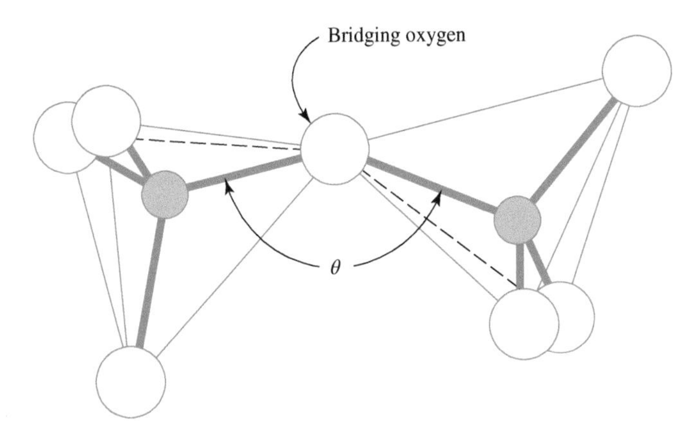

Figure 8.3 Adjacent tetrahedra in silicate crystals or glasses are linked together by a "bridging oxygen." The flexibility of this linkage allows for a wide variety of structural arrangements from the orderly low-temperature quartz crystal (Figure 3.6) to the high-temperature cristobalite (Figure 8.2a) to the random network of silica glass (Figure 8.2b).

increasingly lower costs has only strengthened this field in the ensuing decades. Striking images of atomic-scale structure such as Figure 8.2 are not only works of visual beauty, but they are also scientifically rigorous images of atomic arrangements that are difficult to produce by direct microscopic observation. Computed images of atomic-scale structures have been especially useful in glass science, as experimental tools are limited when inspecting glasses. On the other hand, the regular and repeating arrangements of atoms in crystals have been observable by the use of X-ray scattering (diffraction) methods for more than a century and with the complementary use of electron and neutron scattering for the past several decades.

A few years after the brilliant paper by Zachariasen, a gifted young physicist at MIT, Bertram E. Warren, provided experimental confirmation of the Zachariasen model by applying a sophisticated mathematical analysis to an X-ray diffraction experiment on silica glass. X-ray diffraction patterns are the result of an X-ray beam being scattered by the atoms in the material under inspection. When the atomic arrangement is regular and repeating (crystalline), the pattern tends to be relatively sharp (many strong scattered beams measured at various specific angles). When the atomic arrangement is random (noncrystalline), the pattern tends to be fuzzy (the scattered X-ray intensity changes gradually over a wide range of angles). Figure 8.4 illustrates the distinctive difference between crystalline and noncrystalline X-ray diffraction patterns. The beauty of Warren's mathematical approach is that he used a Fourier analysis that allowed him to calculate the local atomic neighborhood that would correspond to the fuzzy experimental pattern, and the specific result in the case of silica glass was to show that indeed four oxygen atoms were arranged symmetrically around the central silicon precisely, as shown in Figure 8.1. Since that time, the defining characteristic of glass for scientists has been that it demonstrates this **short-range order** of the building block in spite of the fact that its long-range structure is random.

As a practical matter, Warren's original triumph in verifying Zachariasen's claim of the silica tetrahedron as the building block in silica glass was limited to that important fact. The experimental capability of X-ray diffraction in 1936 did not allow him to measure the angles at which the silica tetrahedra are linked (the θ in Figure 8.3). Warren was nonetheless persistent and returned to this classic problem at the end of his career publishing a more refined X-ray diffraction study of the structure of silica glass in 1969 with his final graduate student, Robert Mozzi. They were able to show that the **bond angle** ($\theta_{Si-O-Si}$ in Figure 8.3) varied continuously over a wide range consistent with Zachariasen's random network theory.

And, so, with Warren's experimental verifications, the Zachariasen model has persisted to this day as a fully adequate (and elegant as well as *beautiful*) description of the nature of the atomic structure of pure silica glass. Unfortunately, pure silica glass is expensive, not because of the raw material (clean and pure beach sand will do) but because pure silica (SiO_2) has a very

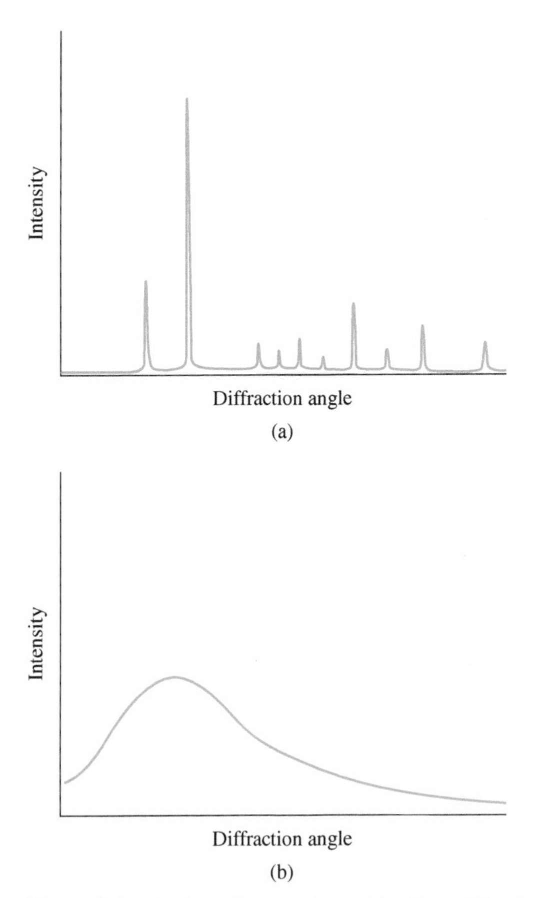

Figure 8.4 A schematic comparison of the X-ray diffraction patterns for (a) a crystalline material and (b) a noncrystalline material of the same composition. The sharp peaks in the crystalline pattern are the result of the long-range order of the atomic structure. The relatively "fuzzy" pattern for the noncrystalline material is the result of the absence of long-range order.

high melting point (over 1700 °C). The cost of melting this material is great and is the first important step in forming the glass. Some applications justify this expense. Pure silica glass crucibles for the production of high-purity silicon crystals in the semiconductor industry are prime examples. In Chapter 3, however, we have acknowledged that common glass products such as glass bottles and stemware for the wine industry have substantial amounts of soda ash (Na_2CO_3) and lime (CaO) in addition to SiO_2 from local sand deposits. Again, it was Bertram Warren who helped us understand the structural role of these **modifier oxides** such as Na_2O that remains after the CO_2 is burned off from the Na_2CO_3 during the glassmaking process.

In an article published in the *Journal of the American Ceramic Society* in 1941, Warren showed how the sodium atoms that are part of the Na_2O in

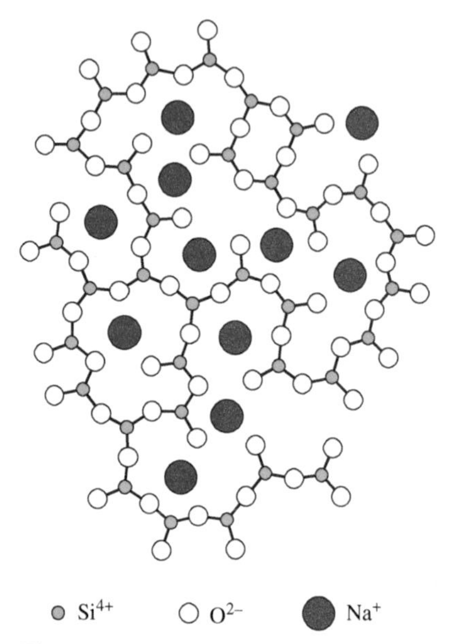

Figure 8.5 The sodium ions in this schematic illustration of a Na_2O—SiO_2 glass break up the linkage of strong Si—O bonds in the pure silica glass (Figure 3.5b) thereby weakening the structure and making the "modified" glass workable at considerably lower temperatures. (Warren, 1941. Reproduced with permission of John Wiley & Sons, Inc.)

Na_2O—SiO_2 glass tend to break up the strong Si—O linkages in the pure silica glass effectively weakening the rigid network of pure silica. Figure 8.5 shows that the considerably weaker Na—O bond gives meaning to the term "modifier" for Na_2O.

In this commentary on atomic bonding, we also see an example of the mantra of materials science and engineering: structure (atomic in this case) leads to properties (strength of the random network), the concept raised in Chapter 1. This weakening of the network structure by adding modifiers such as Na_2O leads to the practical result that the glass can be worked at considerably lower (and more economical) temperatures.

At this point, we again refer the reader to Appendix A to re-emphasize the fact that atomic bonding in silicate glasses has a substantial ionic character. This is especially true for the Na—O bond, with the Na "atom" better characterized as a positive Na^+ ion (or **cation**, the name for a positively charged ion).

Figure 8.6 shows a computer-generated image of the actual three-dimensional structure of a Na_2O—SiO_2 glass, in contrast to the Warren's two-dimensional schematic of Figure 8.5. It is important to note in the case of both pure silica glass (Figure 8.2b) and Na_2O—SiO_2 glass (Figure 8.6) that the random linkage is characteristic of a liquid form as well as the glass. Herein lies the need for some further vocabulary terms and digging deeper into the nature of glass itself.

Figure 8.6 Computer-generated image of the linkage of silica tetrahedra in a Na_2O—SiO_2 glass. This is a three-dimensional version of the schematic in Figure 8.5. In addition, this image of a Na_2O—SiO_2 glass is comparable to that shown in Figure 1.9 except that the "stick-and-ball" configuration of Si—O bonds is replaced by linked tetrahedra representing the cluster of four oxygens around each silicon. (Reproduced with kind permission of Sabyasachi Sen, Department of Materials Science and Engineering, University of California, Davis, CA.)

Figure 8.7 is a plot of volume versus temperature over a wide range for a typical glass-forming material, such as Na_2O—SiO_2. This simple, experimental plot gives us substantial insight into the nature of glass, along with some new vocabulary. We notice that, in general, the volume of the various states of matter

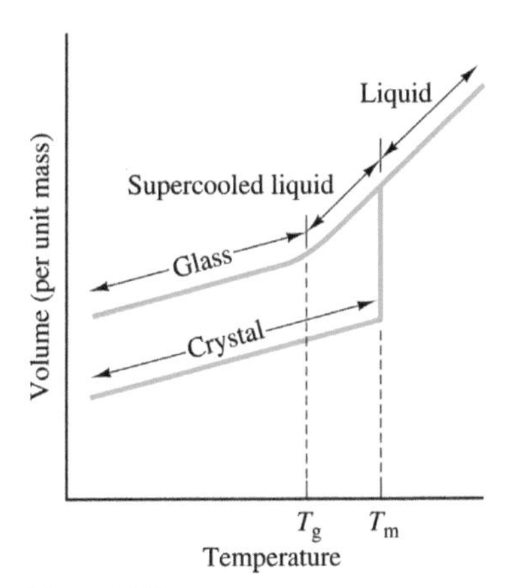

Figure 8.7 A plot of volume versus temperature for a typical glass-forming material. The crystalline form of the material expands steadily upon heating (thermal expansion) but has an abrupt increase in volume upon melting, as the material goes from the relatively efficient atomic packing of the crystal to the more disordered and inefficient packing of the noncrystalline liquid. Upon cooling down from the liquid state, the material can retrace the melting path (crystallize) or, if sufficient time is not allowed for the crystallization to take place, the liquid structure extends below the melting point (supercooled liquid), and eventually the random, liquid-like structure is frozen to form a glass. (Shackelford, 2015. Reproduced with permission of Pearson.)

(crystal, liquid, etc.) increases steadily as we increase temperature. This so-called **thermal expansion** is the result of atomic bonds between adjacent atoms increasing slightly with increasing temperature. Over a century ago, physicists came to understand that atoms are in a constant state of thermal agitation and the extent of such atomic vibration increases steadily with increasing temperature.

We see this thermal expansion for the crystal form of the material to be represented by the slope of the "crystal" portion of Figure 8.7. We also note that there is an abrupt increase in volume at the melting temperature T_m. This distinctive break in the plot of volume versus temperature is the result of going from the relatively efficient packing of atoms in an ordered crystal to the relatively inefficient packing of atoms in a noncrystalline liquid. We can also notice a distinctly higher slope (greater degree of thermal expansion) as the liquid is heated further. This greater slope is associated with the greater mobility of the atoms in the disordered liquid.

Upon cooling down the liquid, there are two possible paths through the volume versus temperature plot of Figure 8.7. One path is to simply retrace the melting path, that is, a thermal "contraction" going down the relatively steep "liquid" portion of the plot followed by a sharp decrease in volume at the melting point and, finally, further thermal contraction going down the relatively shallow "crystal" portion of the plot.

The alternative path upon cooling the liquid gives us the promised, new vocabulary for glass. This alternative is the result of the fact that time is required to rearrange atoms in a material, and this is especially true for silicate glasses. The strong chemical bonds between silicon and oxygen atoms and, to a lesser extent, between sodium and oxygen atoms mean that it is quite possible to outrace the dramatic ordering of atoms required to turn a liquid into a crystal. So, when cooling is relatively rapid, the thermal contraction of the liquid goes seamlessly into the "supercooled liquid" region of the plot, with this new term expanding our vocabulary. An even more novel turn of events is seen at the **glass transition temperature** T_g, where there is a gradual change in slope from the steeper contraction of the liquid states (the "true" liquid and the supercooled one) to the more gradual contraction of the "glass" portion of the plot. So after considerable discussion of glass in the book up to this point, we now have a formal definition of the material in the broader context of glassmaking. It is the solid state in which a liquid-like noncrystalline atomic structure is frozen in place. Both the "crystal" and the "glass" are true, rigid solids with similar thermal expansion slopes. They are separated by the glassmaking operation of melting a crystal and then cooling rapidly enough to freeze in the liquid-like atomic arrangement (such as Figure 8.6).

While Figure 8.7 has given us additional vocabulary including a formal definition of glass, it is somewhat theoretical. We have to recall our earlier descriptions of glassmaking starting in Chapter 1 in which we generally make glass not by simply melting a crystal of the final composition but batching together components (SiO_2, Na_2CO_3, and CaO, etc.).

We can also recall Figure 7.5 in which we described the carefully choreographed dance that a skilled artisan creates in forming a piece of fine stemware. That "dance" is along the viscosity versus temperature curve of Figure 7.4. Now, we see that that dance also takes place between the "glass and supercooled" liquid regions of Figure 8.7, as the material is moved back and forth between highly workable and ultimately rigid states.

Finally, in Figure 8.8, we can admire the beauty of a scientifically rigorous, computer-generated three-dimensional image of Na_2O—CaO—SiO_2 glass at the atomic scale, with its complex intertwining of silica tetrahedra and a random array of Na^+ and Ca^{2+} ions occasionally breaking up the linkages of tetrahedra in their role as "modifiers." While considerable effort has been expended in recent decades by glass scientists in determining if these modifier ions are truly scattered about randomly or if there is some **medium-range order** in their placement, considering them to be randomly distributed is sufficient for our purposes and is a good first approximation to the reality of this structure. In any case, we can acknowledge ourselves to be the beneficiaries of the ability of glass scientists to realistically describe a typical soda-lime-silica glass at the atomic scale in a way that still eludes experimentalists. As we pointed out at the outset of this book in Chapter 1, the underlying principle of materials science in general and glass science in particular is that "structure leads to properties." So, the beautiful understanding of structure in this chapter will allow us to appreciate the mechanical and optical behavior of glasses as we move on to the next two chapters.

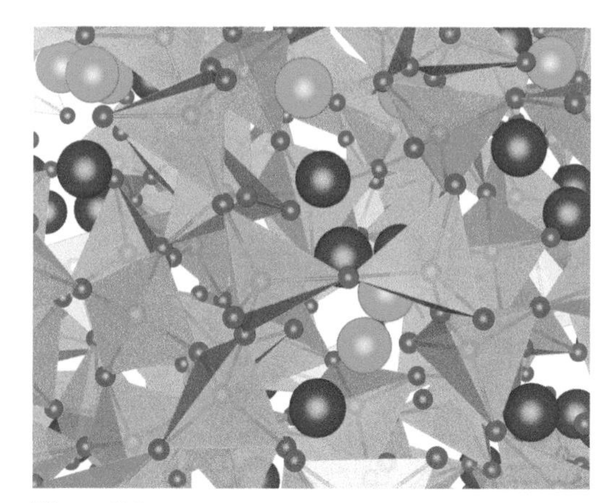

Figure 8.8 Computer-generated image of the linkage of silica tetrahedra in a Na_2O—CaO—SiO_2 glass, with Na^+ and Ca^{2+} ions serving as "modifiers" by breaking up the tetrahedral linkages. The small yellow spheres represent silicons; the small red spheres represent oxygens; the large purple spheres represent the Na^+ ions; and, the large green spheres represent the Ca^{2+} ions. Note that the light yellow, translucent tetrahedra show the SiO_4 clusters of Figure 8.1. (Reproduced with kind permission of Jincheng Du, Department of Materials Science and Engineering, University of North Texas, Denton, TX.)

BIBLIOGRAPHY

Allen, Samuel and Edwin Thomas, *The Structure of Materials*, John Wiley & Sons Inc., New York (1999).

Davila, Lilian, Subhash Risbud, and James Shackelford, "Interstitial Nanostructures in Engineered Silicates," *Ceramic Transactions*, **137**, 209–219 (2003).

Kingery, W. David, H. Kent Bowen, and Donald Uhlmann, *Introduction to Ceramics*, 2nd Edition, John Wiley & Sons Inc., New York (1976).

Shackelford, James, *Introduction to Materials Science for Engineers*, 8th Edition, Pearson, Upper Saddle River, NJ (2015).

Chapter 9

The Heel of Achilles – Why Glass Breaks

"Don't tell me the moon is shining; show me the glint of light on broken glass."

Anton Chekhov, 1886, as quoted in *The Quotable Book Lover*, Ben Jacobs and Helena Hjalmarsson, Eds., Skyhorse Publishing, New York (1999).

Broken glass is such a part of our human experience that a great writer such as Anton Chekhov can use it as a literary device – an everyday lens to reveal the moon's reflected light on a clear night. The propensity of glass to break is of special importance to the wine industry, given the material's role in transporting, storing, and consuming wine. And, so we need to spend some time considering this Achilles heel for a material that seems so ideal in so many ways, save one – the all too likely fate should we not handle it with care.

Imagine the following scene in a kitchen after dinner, when you are cleaning up after a meal. You are drying a metal pan and, perhaps due to a bit too much wine, it slips from your hand and falls on the floor. Aside from the clang of the collision with the floor, as it bounces once (perhaps twice), it is by and large undamaged; perhaps there is a slight dent. But what if your clumsiness occurred not while drying the pan but while drying your favorite stemware? The crashing sound is not a clang but a shattering. Get a broom and watch your step (Figure 9.1)!

We doubt if many readers have not done these "experiments" at some point over the years. The metal pots and pans tend to bounce with little damage. Drinking glasses rarely survive, especially thin and elegant wine glasses. To appreciate the reason for this distinctly different behavior for these two different materials, let us once again repeat the mantra of materials science: *structure*

The Glass of Wine: The Science, Technology, and Art of Glassware for Transporting and Enjoying Wine, First Edition. James F. Shackelford and Penelope L. Shackelford.

Figure 9.1 A tale of two materials: a metal pan that survived a fall onto a kitchen floor shows a negligible effect, while a wine glass displays a catastrophic result.

leads to properties. In this case, the structural basis of breaking versus bouncing is understood first of all at the atomic level.

Metals tend to be **ductile**, that is, they can be deformed substantially without breaking. Again, imagine a simple mechanical experiment. You hold an empty aluminum can in your hand. You can squeeze it and fairly easily produce a permanent deformation, literally crushing it in your hand. The basis of this *ductility* or deformability is again atomic-scale structure. Common metal alloys typically have relatively simple crystalline structures that allow layers of atoms to literally slide past each other in response to an applied mechanical load. The details of these so-called *slip systems* are outside the scope of this book on glass, but Figure 9.2 gives a simple schematic illustration of the concept. (Readers with some familiarity with physical metallurgy will recall that the atomic scale explanation of this slippage of layers of atoms past each other involves structural defects called "dislocations." Anyone interested in a review of those details can find them in an introductory materials science text such as those in the bibliography of this chapter.)

On the other hand, ceramics with their chemistry similar to glasses but with a regular crystalline arrangement of atoms are **brittle** materials. Why? Again, the answer lies in the atomic structure. The crystalline arrangement of atoms in ceramics is considerably more complex than in common metals. The result is that the "slip" illustrated by Figure 9.2 is not possible in common ceramics, and we can all generally recall an "experiment" of dropping a ceramic plate on the kitchen floor with disastrous results. For glasses, completely lacking in crystallographic order, such slip is even more out of the question. And, so the relatively strong bonding of the random network of atomic bonds described by Zachariasen leaves us with a very brittle and highly breakable material.

With a sense of the role of atomic-level structure on the nature of fracture in glass, we can now turn to the macroscopic scale at which engineering

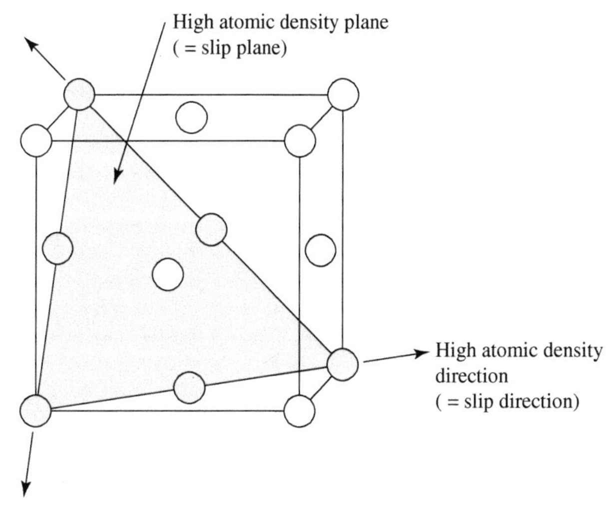

Figure 9.2 A schematic illustration of the relatively easy mechanism of "slip" of layers of atoms past others at the atomic scale in a metal. In this example, the multiple directions available for atomic slip correspond to those for common metals such as aluminum, copper, and some stainless steels. Such "slip systems" are defined as combinations of specific (high atomic density) crystallographic directions and planes.

measurements are made. Figure 9.3 illustrates how most engineering materials are evaluated for strength and ductility. The technique is, in principle, quite simple. The **tensile test** involves pulling a rod of the material with a carefully controlled and continually increasing load.

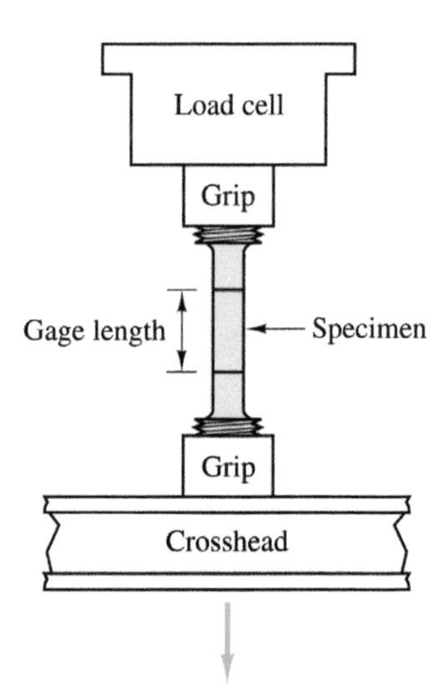

Figure 9.3 A schematic illustration of the tensile test that is used to measure the strength and ductility of common structural materials. (Shackelford, 2015. Reproduced with permission of Pearson.)

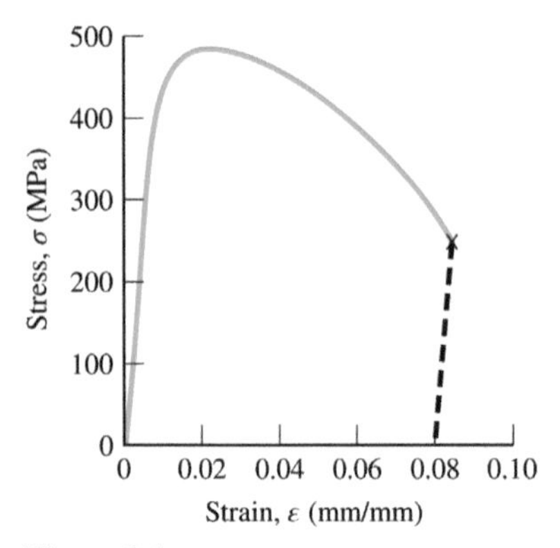

Figure 9.4 A typical result of a tensile test of a structural metal alloy, such as the material that would have been used to manufacture the metal pan in Figure 9.1. (Shackelford, 2015. Reproduced with permission of Pearson.)

The result of the tensile test for metallic materials compared to brittle ceramics and glasses is quite different. If we were to obtain a simple, cylindrical rod of the material used to make the metal pan in Figure 9.1 and subject that rod to the pull test in the experimental system of Figure 9.3, the result would be like the plot in Figure 9.4. This so-called *stress versus strain* plot tells a complex story of the deformation of the metal that results from the gradual loading of the material all the way to ultimate failure.

Before we recount this tale of the gradual destruction of a metal alloy, we need to define the two terms that label the two axes of the plot. *Stress* is simply the mechanical load applied to the metal rod by the tensile test machine divided by the cross-sectional area of the rod. The units of stress are the basic unit of force (1 N) divided by the basic unit of area (1 m^2) that is equivalent to 1 Pa, named in honor of the great French physicist Blaise Pascal. (For those, mostly Americans, who continue to use the units of psi for pressure, we note that 1 Pa = 0.145×10^{-3} psi or more formally 0.145×10^{-3} lb_f/in^2.) Because the forces necessary to pull a metal bar or other structural materials to destruction are quite large, the magnitude of stress is typically expressed in millions of pascals or megapascals.

Strain is then a measure of the extension of the bar in response to the mechanical stress. The strain is specifically defined as the fractional increase in length of some arbitrary section of the metal bar, or the so-called **gage length** – typically a length of 2 in. (about 50 mm).

The complex story of the failure of the metal alloy in Figure 9.4 can be told in two parts. First, we see that the initial stretching of the metal bar is shown as a straight line. Equally important is the fact that, if the load were to be released during any part of this linear region, the plot would retrace that straight line back to the origin (where there is zero strain under a stress of 0 MPa). This straight-line portion of the plot is an example of **elastic** (or temporary) **deformation**. As long as we stay within that region of the plot, we can expect that a load (stress) will lead to some elongation (strain) that will be removed whenever we remove that load. The slope of the straight-line portion is called the **elastic modulus** and, in practical terms, represents the *stiffness* of the material.

The second part of our story involves the highly *nonlinear* curve that evolves as we move beyond the straight-line segment. This region tells the story of **plastic** (or permanent) **deformation**. We can recall the image of the crushed aluminum can invoked at the beginning of this segment. That crumbled shape is permanent; it is there to stay. A metallurgist can catalog the stages of this plastic deformation by defining the markers along the plot of Figure 9.4. The **yield strength (YS)** is roughly defined as the line of demarcation between elastic deformation and plastic deformation. (It is, in fact, slightly into the plastic region, defined by a conveniently measured offset of 0.2% deformation.) The **tensile strength (TS)** is the highest value of stress measured over the entire plastic range. The drop-off in stress beyond TS does not imply the material is becoming weaker, rather it is a simple manifestation of the fact that the rod being pulled in tension is deforming rapidly and the cross-sectional area is substantially smaller than that at the outset. (Note again that the **engineering stress** is defined relative to the *initial* cross-sectional area, not the true value that is shrinking rapidly just before failure.) And, at the final point of failure (X marks the spot in Figure 9.4), the metal rod snaps, with a small amount of elastic snap back, that leads to the value of ε_f or "strain at failure" on the strain axis that serves as our formal definition of **ductility**.

Now for the mechanical behavior of a brittle glass pulled in tension in an experiment comparable to that shown in Figure 9.3. The austere result for brittle glass is shown in Figure 9.5. In this case, there is only elastic deformation – with **brittle fracture** occurring at the end of the straight-line segment.

We now see the consequence of the lack of "slip systems" in glass – the absence of a structural basis for atomic-scale sliding of atomic planes past each other. That "slip" gave us the second part of the deformation story in Figure 9.4, namely, plastic deformation. Figure 9.5 tells only one story – elastic deformation that ends in brittle fracture.

But, wait, it gets worse! Thanks to an aeronautical engineer who studied the failure of materials a century earlier, we now are aware that small defects can amplify the problem. The British engineer Alan Arnold Griffith (1893–1963) was an important contributor to aeronautics and among the first to suggest that gas turbines would be feasible propulsion systems. His fascination with engineered materials led him to publish an important paper in 1920 that analyzed the strength of glass fibers. The result of his experiments led to a simple geometrical description of the unavoidable surface cracks that will be produced in such materials as a result of manufacturing and subsequent handling. Griffith illustrated these flaws as elliptical cracks (Figure 9.6) and the now famous equation that

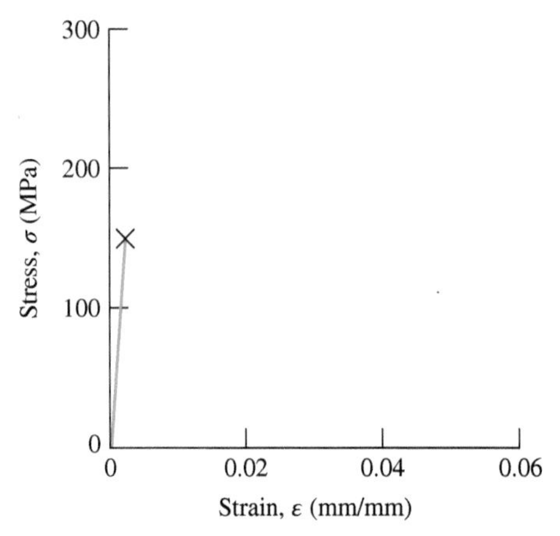

Figure 9.5 A typical result of a tensile test of common glass, such as the material that would have been used to manufacture the stemware in Figure 9.1.

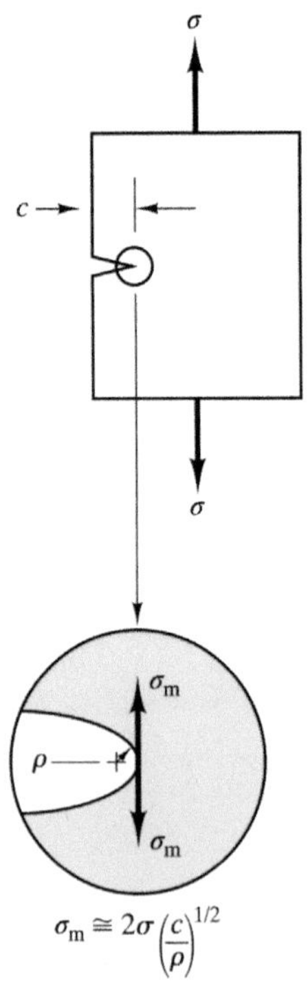

Figure 9.6 The stress at the tip of a Griffith crack, σ_m, is substantially higher than the nominal stress on the overall glass, σ, by a factor given by the Griffith equation (Equation 9.1). (Shackelford, 2015. Reproduced with permission of Pearson.)

bears his name indicates that the stress at the crack tip, σ_m, is substantially larger than the overall value applied to the glass bar as a whole, σ:

$$\sigma_m \approx 2\sigma(c/\rho)^{1/2} \tag{9.1}$$

where c is the depth of the surface crack and ρ is the radius at the crack tip. The "stress amplification" identified by the Griffith equation is 2 $(c/\rho)^{1/2}$, a rather large number as crack depths are typically in the micrometer (μm) range and the crack tip can be quite sharp with a radius on the order of an atomic dimension (less than one nm with 1000 nm = 1 μm). And, so, thanks to Griffith, we can refer to these inevitable, small scratches as "stress risers," defects that make typical glass vessels weak as well as brittle. While the glass may have a reasonably high strength if produced in a flaw-free form, the Achilles Heel of small surface scratches cause the actual breaking strength to be exceeded when the nominal stress load might seem to be quite modest.

With the substantial difference in mechanical behavior of metals and glasses now quantified, we can compare side by side the values of their properties. Before comparison however, we should acknowledge that brittle ceramics and glasses tend not to be tested in the tensile test of Figure 9.3. As a practical matter, gripping these brittle materials in that configuration is difficult and results tend to be rather scattered. Instead, a bending test as shown in Figure 9.7 is more convenient and reproducible.

Not only is the test for brittle glass different in geometry than the tensile test for metal alloys, the property measured is also somewhat different. Equation 9.2 defines the **flexural strength (FS)**, sometimes called the *modulus of rupture*, that is calculated for the bending test:

$$\text{FS} = (3FL)/(2bh^2) \tag{9.2}$$

where F is the applied force and b, h, and L are the dimensions of the test piece, as defined in Figure 9.7. Fortunately, the value of flexural strength can be compared meaningfully with the breaking stress in a tensile test, as the outermost

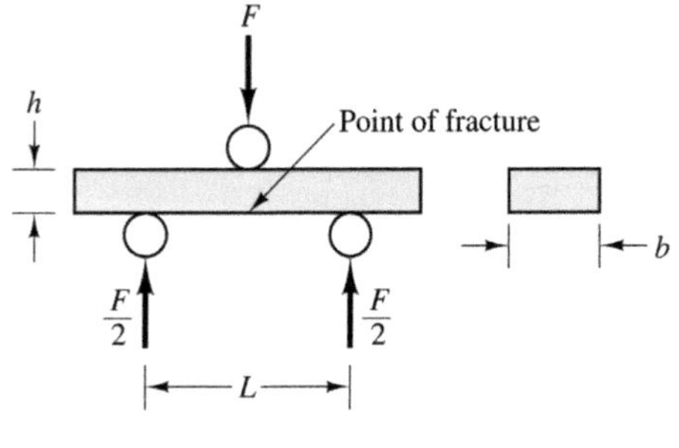

Figure 9.7 The mechanical performance of brittle ceramics and glasses is typically measured in a bending test. (Shackelford, 2015. Reproduced with permission of Pearson.)

Table 9.1 Modulus of Elasticity and Strength for Some Common Glasses Compared to Typical Ceramics and Metal Alloys

Material	*E* (GPa)	Flexural strength (MPa)	Tensile strength (MPa)
Glasses			
Silica glass	73	104	
Soda-lime-silicate glass	66	150	
Borosilicate glass	60	69	
Ceramics			
Alumina	380	276–1034	
Mullite (aluminosilicate)	50–220	150–270	
Glass-ceramics (e.g., Corning Ware®)	83–138	70–350	
Metals			
Gray cast irons	66–162		118–1,276
316 stainless steel	193		515–620
Ti-6Al-4V	114		896–1,172

Source: From data collections in J.F. Shackelford *et al.*, *CRC Materials Science and Engineering Handbook*, 4th Edition, CRC Press, Boca Raton, FL, 2016; *Engineered Materials Handbook*, Vol. 4, *Ceramics and Glasses*, ASM International, Materials Park, OH, 1991; and W.D. Kingery, H.K. Bowen, and D.R. Uhlmann, *Introduction to Ceramics*, 2nd Edition, John Wiley & Sons, Inc., New York, 1976.

edge of the test sample in the bending test is under tension. So now we can move ahead to the "side-by-side" comparison of metals and glasses as summarized in Table 9.1. In reflecting on the various data in Table 9.1, we can note that we are basically focusing on the two most basic characteristics of a structural material – its strength and stiffness. As just noted, the flexural strength and tensile strength are comparable terms, although they are associated with materials breaking in very different ways (Figure 9.4 versus Figure 9.5).

As for the elastic modulus, we can again invoke our "structure leads to properties" mantra. From the structural perspective, the modulus is a direct reflection of the strength of the chemical bond between adjacent atoms. So, for a glass, it indicates an average of the relatively strong Si—O bond and the somewhat lower strength bonds between Na—O and Ca—O. In steel, the relevant bonds are those between adjacent Fe atoms, and others such as Fe—Cr and Fe—Ni in stainless steel. Then the manifestation of this interatomic bonding on the engineering scale is, as we stated earlier, the *stiffness* of the material, that is, the straight-line portion of the stress versus strain curve (the entire curve for glass (Figure 9.5) and the first part of the curve for a metal alloy (Figure 9.4)).

With our definitions of strength and stiffness in hand, we can now appreciate the relative performance of the three broad categories of materials in Table 9.1. For the entire table, we see strength values varying widely from less than 100 MPa to over 1000 MPa. The strength of the common glasses generally runs below that of the typical ceramics and metals. Nonetheless, the strength is adequate for normal use; one simply needs to maintain constant vigilance in regard to avoiding the natural tendency toward brittle fracture. The modulus or stiffness of glass is also somewhat smaller than most ceramics and metals, but again the values are quite adequate for practical use.

To summarize, we see that glass is a material with good strength and stiffness and clearly adequate for its well-established role of storing, shipping, and consuming wine, but we must be eternally vigilant armed with our knowledge of its Achilles heel of breaking in a catastrophic and brittle fashion with its strength compromised by the inevitable, minor scratches that come from the manufacturing and handling of this material. Yes, it is clearly adequate, but perhaps its greatest attribute (property) is that it is also *clear*. We now move to a discussion of why glass not only contains our wine but also provides a window to it.

BIBLIOGRAPHY

Callister, William and David Rethwisch, *Materials Science and Engineering: An Introduction*, 9th Edition, John Wiley & Sons, Inc., New York (2014).

Kingery, W. David, H. Kent Bowen, and Donald Uhlmann, *Introduction to Ceramics*, 2nd Edition, John Wiley & Sons, Inc., New York (1976).

Shackelford, James, *Introduction to Materials Science for Engineers*, 8th Edition, Pearson, Upper Saddle River, NJ (2015).

Chapter 10

Let It Be Perfectly Clear – Why Glass Is Transparent

"A spring there is, whose silver waters show
Clear as a glass the shining sands below."

Alexander Pope in his translations from Ovid (Sappho to Phaon), lines 179–180

While the story of mechanical behavior was a cautionary tale, the story of optical behavior in glass is, well, *as clear as glass!* For many materials, their optical behavior is more important than the mechanical. That is especially true for the drinking glasses from which we enjoy wine. In Chapter 12, we will see that modern stemware has evolved into an elegantly simple design dependent on transparency to provide an unhindered view of the liquid within. Some wine bottles to be discussed in Chapter 11, however, depend on shielding their contents from ambient light that could adversely affect final taste. The transmission, reflection, and absorption of light are all important properties that we must explore in this chapter.

We begin by asking: What is light? Figure 10.1 gives an especially fundamental definition, namely, it is an electromagnetic oscillation with wavelength, λ, and a velocity c, the speed of light.

Perhaps the more relevant question is "what is visible light?" The answer to that question can be found in Figure 10.2. The **electromagnetic spectrum** covers a very wide range of frequencies, from the high frequencies of X- and gamma-rays to the low frequencies of radio waves. The various categories along this wide spectrum are associated with different sources of the electromagnetic vibrations. For example, X-rays are produced by high-voltage machines and gamma-rays by the radioactive decay of certain isotopes. The word "light" is

The Glass of Wine: The Science, Technology, and Art of Glassware for Transporting and Enjoying Wine, First Edition. James F. Shackelford and Penelope L. Shackelford.

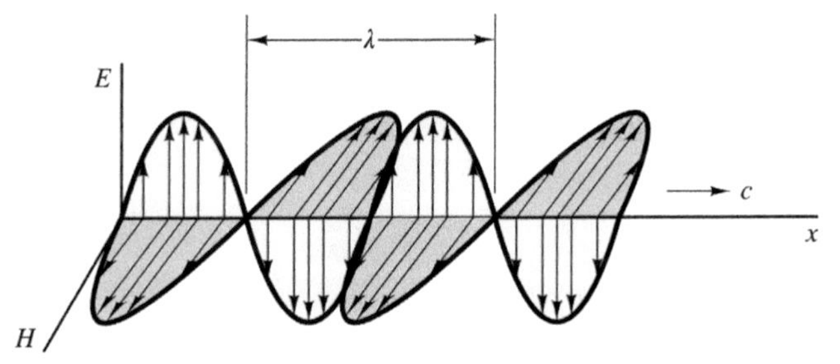

Figure 10.1 Light corresponds to the sinusoidal oscillation of both an electrical field (E) and a magnetic field (H). These two oscillations occur in perpendicular planes. The wavelength is given by λ, while the wave travels at the speed of light, c. (Shackelford, 2015. Reproduced with permission of Pearson.)

often used in conjunction with these many different forms of radiation. On the other hand, **visible light** that we can perceive with our eyes is a narrow slice of the overall spectrum, a range of approximately 400–700 nm, wedged between the lower frequency ultraviolet radiation and the higher frequency infrared radiation.

We can begin this exploration of the nature of light by acknowledging that it is, in fact, of two forms: a particle and a wave. One of the triumphs of twentieth-century physics was to characterize these simultaneous forms (the wave particle duality). Each wavelength of the electromagnetic radiation can be simultaneously described as an equivalent particle with an energy E given by

$$E = h\nu \tag{10.1}$$

where h is the Planck's constant ($=0.6626 \times 10^{-33}$ J·s) and ν is the radiation frequency, that is, in turn, $=c/\lambda$, where c is the speed of light ($=0.2998 \times 10^9$ m/s) and λ is the wavelength of the radiation. This basic relationship between frequency and wavelength ($\nu = \text{velocity}/\lambda$) is true of any waveform, from ocean

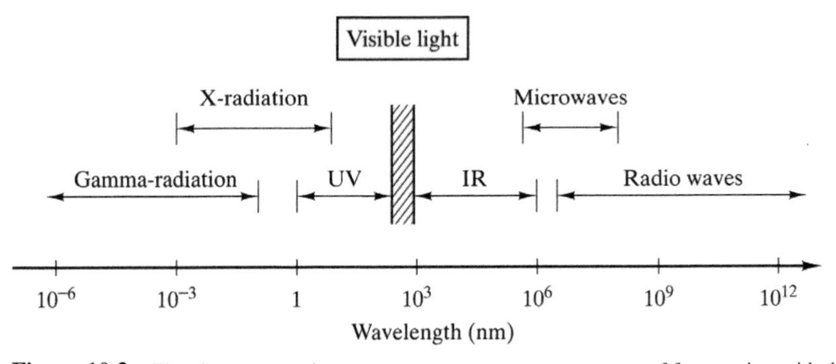

Figure 10.2 The electromagnetic spectrum covers an enormous range of frequencies, with the visible light spectrum being a relatively narrow slice between approximately 400 and 700 nm. (Shackelford, 2015. Reproduced with permission of Pearson.)

waves to X-rays, but the exceptionally high-velocity term, the speed of light, c, is unique to the electromagnetic spectrum. Combining this velocity definition with Equation 10.1 gives an important relationship between energy and wavelength:

$$E = h\nu = hc/\lambda \tag{10.2}$$

So, energy is greater for smaller wavelengths. Again referring to Figure 10.2, we can appreciate that the very short wavelengths of X- and gamma-rays correspond to their reputation as "hard" or "penetrating" radiation. Medical and dental technicians discretely step out of the room or behind shielding when they are taking medical and dental X-rays for precisely this reason. While visible light is "softer" by orders of magnitude, we are still admonished to take care from overexposure to sunlight, as it can do substantial skin damage (although we are also frequently told correctly that it is the "harder" ultraviolet radiation that is the main culprit).

Before continuing with our discussion of the properties of light, we should emphasize an important, general characteristic – light rays (photons) tend to be absorbed by electrons, at least by *certain* electrons. This important fact, in turn, returns us to our consideration of atomic bonding and the associated discussion in Appendix A. As noted there and earlier in the book, metals are good electrical conductors, while typical ceramics, glasses, and polymers are electrical insulators. The good conductivity of metals is the result of its atomic bonding. As detailed in Appendix A, metallic bonding involves the equivalent of a "gas" of electrons, freely migrating electrons that can quickly move through the material under a voltage gradient. On the other hand, ceramics and glasses involve substantial ionic bonding in which electrons are transferred from some atoms to others, leaving behind positively charged cations (such as Si^{4+}) and producing negatively charged **anions** (such as O^{2-}). The coulombic bond between the cations and anions provides the "glue" holding the material together. Similarly, polymers involve substantial covalent bonding, in which bonding electrons are shared between adjacent atoms (often carbon atoms). For both ionic and covalent bonding, the bonding electrons are "spoken for" in that they are tightly bound to adjacent nuclei.

On the other hand, the bonding electrons in metals and their alloys are "unbound" and free to roam throughout the material. These free electrons that provide electrical conductivity are precisely the ones that can interact with incoming photons and absorb them. The result is the shiny and opaque metals with which we are familiar. Light does not penetrate but is reflected off the surface.

So, it is the ionically and covalently bonded ceramics, glasses, and polymers that can all be "clear as glass." With this basis of transparency in hand, we can now discuss the most fundamental optical property of glasses that plays an important role in their use, namely, the **index of refraction**.

The index of refraction allows us to quantify the way in which light passes through a transparent material as well as how it is reflected off the surface. We

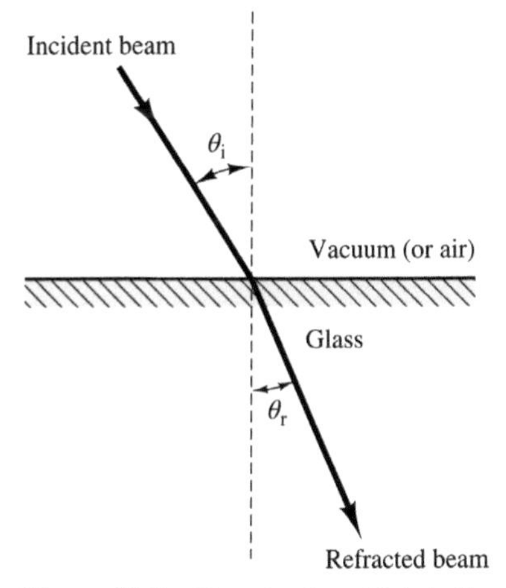

Figure 10.3 The refraction of light is illustrated as a beam of light passes from a vacuum (or air) into a transparent glass. (Shackelford, 2015. Reproduced with permission of Pearson.)

implied how important this property is in Chapter 3 when we were critiquing the use of the term "lead crystal" as associated with fine stemware. There we noted that the distinctive crystal-like sparkle of such glass is the result of the high index of refraction provided by the lead content. We also saw in Figure 6.5 how the refraction of light through grape juice allows the °Brix to be determined by a refractometer and thus guide the winemaker in choosing the optimal harvest date.

So, what is the refractive index? It is simply the ratio of the speed of light in a vacuum to that in the transparent medium. It is simultaneously equal to the ratio of the sine of the angle of incidence to the sine of the angle of refraction within the transparent medium. These equivalences are summarized in Figure 10.3 and Equation 10.3 defines the index of refraction, n, as

$$n = c_{\text{vac}}/c = \sin\theta_{\text{i}}/\sin\theta_{\text{r}} \tag{10.3}$$

where c_{vac} is the speed of light in a vacuum (essentially equal to that in air) and c is the speed of light in the glass and θ_{i} is the angle of incidence and θ_{r} is the angle of **refraction** as defined in Figure 10.3.

Typical values of n fall between 1.5 and 2.5, with most silicate glasses having values close to 1.5. As noted in Chapter 3, lead has a high index of refraction. When it is added to glass in the form of lead oxide (PbO), the average value of n for the glass is correspondingly increased, accounting for the high degree of internal reflection and subsequent "sparkle" that reminds us of fine gemstones such as diamond. Of course, these various values of n well above 1 tell us that light slows down significantly within the glass ($c_{\text{vac}} > c$).

So, now we have a sense of how and why light refraction through glass plays an important part in our appreciation of the material *and* the wine

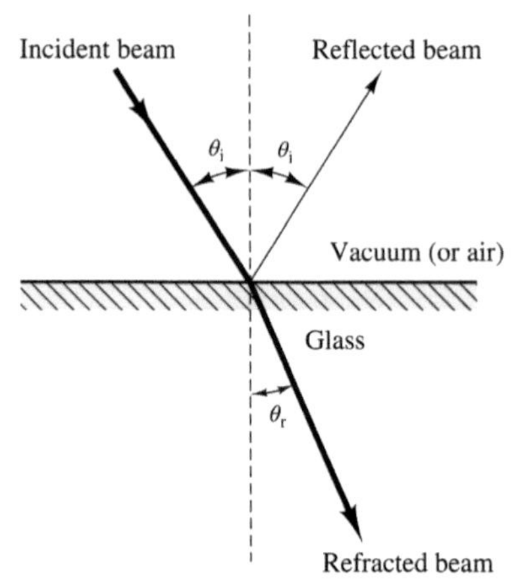

Figure 10.4 Light reflection occurs at the surface of a transparent glass (at an angle equal to the angle of incidence), along with refraction. (Shackelford, 2015. Reproduced with permission of Pearson.)

contained within. This property also tells us something about how light is reflected at the surface. Before we quantify that relationship, we can state one of the simplest findings from the science of optics: the angle of **reflection** equals the angle of incidence, as illustrated in Figure 10.4 and given by the equation

$$\theta_{\text{reflection}} = \theta_{\text{i}} \qquad (10.4)$$

Beyond that useful and elegantly simple fact, the index of refraction gives us the additional information about what fraction of the beam intensity is reflected back from the surface. This fraction is called the reflectance R and is given by Fresnel's formula named in honor of the great nineteenth-century French physicist Augustin Jean Fresnel and is but one of his many contributions to the theory of light:

$$R = ([n - 1]/[n + 1])^2 \qquad (10.5)$$

While Equation 10.5 is strictly valid for normal incidence where $\theta_{\text{i}} = 0°$, it is a good approximation over a wide range of θ_{i}. The nature of Equation 10.5 indicates that high n value glasses will be more highly reflective. For example, the value of R for a lead silicate glass with $n = 2.5$ is 0.429 (nearly half of the light is reflected), while the R for a common lead-free silicate glass with $n = 1.5$ is 0.200 (less than a quarter of the light is reflected).

While useful, Figure 10.4 is somewhat idealistic in that it assumes the glass surface is perfectly smooth, and a thin beam of light will be reflected and refracted at precise angles. In reality, surfaces are not always perfectly smooth. In some cases as illustrated in Figure 10.5, substantial variations can occur. The

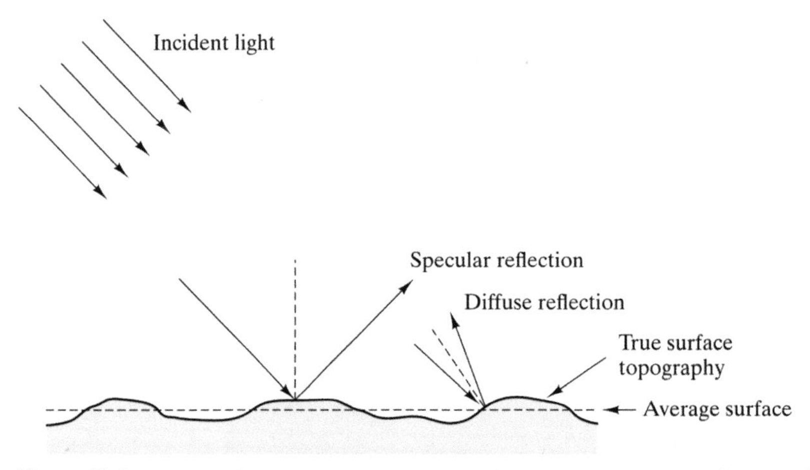

Figure 10.5 Reflection is of two types: specular reflection relative to the "average" or smooth surface and diffuse reflection from local areas of surface roughness. (Shackelford, 2015. Reproduced with permission of Pearson.)

result is that the reflection off an irregular surface will include the reflection of some light rays from those parts parallel to the average or ideal surface, while some light rays reflect off parts that are not parallel and hence reflect at different overall angles relative to the average. We call the reflection relative to the "average" as **specular reflection** and reflection due to surface roughness as **diffuse reflection**. The relative amount of each type of reflection is nicely illustrated by polar diagrams such as that in Figure 10.6 showing predominantly specular reflection off a relatively smooth surface.

One application of the principles just illustrated is that it helps us appreciate the technology behind the popular use of winery logos on stemware used in tasting rooms (Figure 10.7).

It is also worth noting that Figure 10.6 in conjunction with Equation 10.5 tells us that a high index of refraction in combination with a smooth surface gives

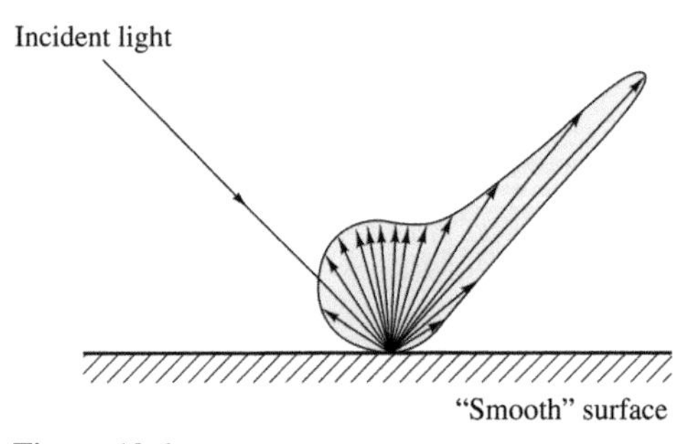

Figure 10.6 A polar diagram showing largely specular reflection from a relatively smooth surface. (Shackelford, 2015. Reproduced with permission of Pearson.)

Figure 10.7 The difference between specular and diffuse reflection is at the heart of the technology of etched labels on wine glasses used in winery tasting rooms. The diffuse reflection from the etched portion of the glass surface is in stark contrast to the specular reflection from the unetched areas. This label is seen at the Quinta da Pacheca in the Duoro Valley of Portugal. (Reproduced with kind permission of Quinta da Pacheca.)

a high level of *surface gloss*. While not so important for wine glasses and bottles, it helps us appreciate the strategy of inexpensive pottery: adding a bit of lead to a surface glaze (glass coating) on the ceramic piece gives the eye-catching shiny surface.

A material that is as clear as glass serves as a working definition of the word **transparent**, a common descriptor for a range of daily experiences, from clear glass to how forthcoming a given politician might be (perhaps more "transparent" in revealing names of campaign donors!). For our understanding of the technology of glass, we can define "transparent" in a formal way: the transmission of a clear image.

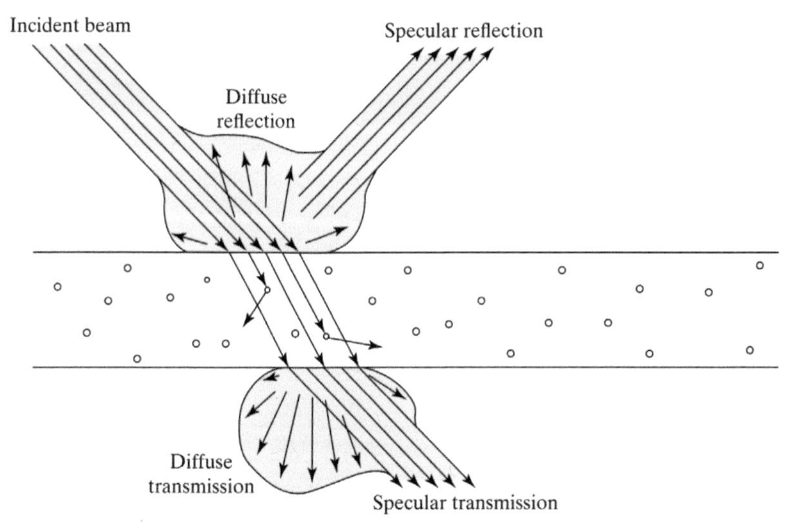

Figure 10.8 Translucency is defined with polar diagrams showing the combination of specular and diffuse reflection and refraction that occur for light transmitted through a slab of glass with a modest amount of porosity. An image through this glass will be "diffuse." (Shackelford, 2015. Reproduced with permission of Pearson.)

Not all glass, however, transmits a clear image. This fact provides two new vocabulary terms of somewhat arbitrary distinction. The transmission of a diffuse image defines **translucent**. The complete loss of image transmission defines **opaque**. A limited number of small pores or small second-phase particles can lead to translucency. Figure 10.8 illustrates translucency with polar diagrams for the case of porosity, and Figure 10.9 shows how, on the microscopic scale, the basic mechanism of image degradation is the refraction of light at the interfaces between the glass and a particular pore. The index of refraction for the pore is essentially the same as that for air or a vacuum, $n = 1$, and the value for the glass will, of course, be considerably greater, $n > 1$, with the exact value depending on the specific glass composition. Opacity occurs when the concentration of pores or second-phase particles is sufficiently large. In fact, image transmission can be completely lost when the concentration of such scattering centers is as small as 3%. It is also worth noting a rule of thumb from optics, namely, light scattering is most efficient when the scattering centers (pores or particles) are approximately the same as the light waves (in the wavelength range of 400–700 nm). So, image degradation will be most pronounced by pores/particles in that size range. In fact, extremely small pores as small as a few nanometers in size can be quite ineffective in scattering light waves, allowing transparency in the presence of porosity.

In general, fine stemware requires nothing less than transparency. On the other hand, wine bottles may have less rigorous demands, with some minor porosity from manufacturing correspondingly tolerable. Of course, the concern

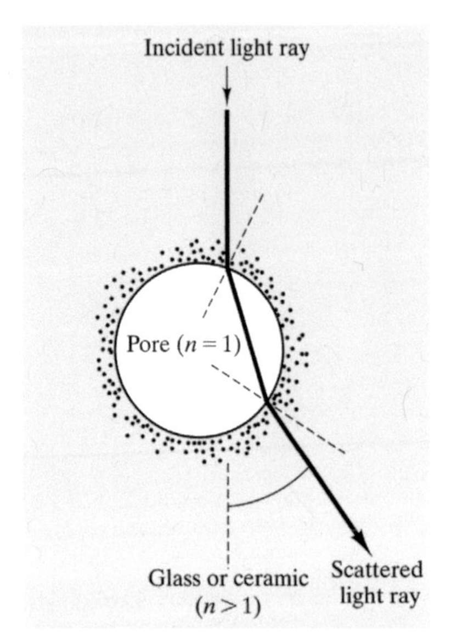

Figure 10.9 The degradation of an image transmitted through a glass such as that shown in Figure 10.8 is due to a multitude of scatterings caused by each of the pores in the material. Here, the refraction of a single light ray by a single pore is illustrated. (Shackelford, 2015. Reproduced with permission of Pearson.)

for these pores serving as stress risers, as introduced in the previous chapter, cannot be overlooked. This concern is especially true for bottles stored under pressure such as Champagne and other sparkling wines. We will return to this issue in Chapter 11.

Before leaving the topic of opacity, we should contrast the scattering mechanism for opacity in an otherwise transparent material such as glass with the characteristic opacity of metals introduced at the beginning of this chapter. For metals, we recall that an otherwise pore-free sample will be quite opaque simply because of the presence of the mobile gas of bonding electrons, free to absorb any and all light rays that try to penetrate the material.

Whether or not porosity might serve to obscure our view of the wine contained within a bottle, we often find bottles of wine that effectively shield light from its contents with the use of color. So, now let us turn to the fascinating topic of **color** in glass.

While opacity in glass was seen to be a scattering mechanism in contrast to the electronic absorption mechanism of opacity in metals, color in glass is another example of light absorption by electrons, but in a more limited way. The coloring of glass was a necessary topic to introduce in Chapter 7 as part of a wide-ranging discussion of glassmaking. With our expanded understanding of optics in this chapter, we can now delve a bit deeper into the science and technology of color.

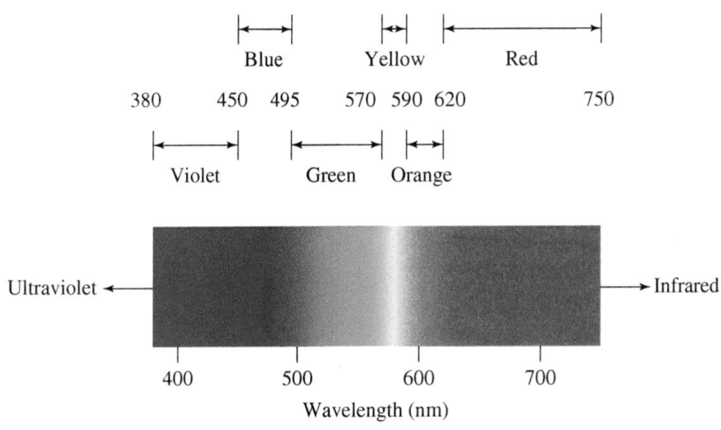

Figure 10.10 As noted in Figure 10.2, the visible light spectrum is a relatively narrow slice of the electromagnetic spectrum with wavelengths between approximately 400 and 700 nm. The familiar color spectrum is associated with this range, with the rainbow of colors spreading from violet at the short wavelength end around 400 nm to red at the long wavelength end around 700 nm.

The range of valence electron energies for the bonding electrons of metals leads to all light rays across the entire visible spectrum (approximately 400–700 nm) being absorbed; hence, complete opacity. In contrast, nonbonding electrons within some *transition metal ions* in glasses can absorb light rays of specific energies, cutting out a mere slice of the overall visible spectrum. To appreciate the details of this electronic mechanism of color production, we need to return to the electromagnetic spectrum introduced in Figure 10.2 but focus on the range of visible light photons. Figure 10.10 shows this range in detail with the corresponding colors associated with narrow bands of wavelengths, from violet around the short wavelength (400 nm) end up to red around the long wavelength (700 nm) end. The bands of individual colors shown in Figure 10.10 indicate that the "boundaries" between colors are diffuse and that there are similarly diffuse gradations below 400 nm between violet and ultraviolet radiation and above 700 nm between red and infrared radiation.

So, we see that a specific wavelength of visible light is associated with a specific color. We can also recall that a specific wavelength is associated with a specific photon energy through Equation 10.2. The consequence of these equivalences is that any specific color is associated with some specific energy. This color–energy relationship is our portal to understanding how we can engineer color into a glass, but passing through this portal will again require us to understand some fundamentals of chemistry.

As with our introduction of Appendix A in Chapter 8 relative to atomic bonding and earlier in this chapter in regard to the absorption of light by the bonding electrons in metals, some may want to turn to Appendix A for an introduction (or refresher) on the details of the planetary model of atomic structure, wherein lies the basis of color formation. There one will see that the bonding electrons for a given atom are those in the outermost orbital around an atomic nucleus, but there are also a number of inner orbitals populated by electrons that do not participate in bonding. Nonetheless, these electrons occupy distinct energy levels, and, in certain transition metal ions, can be promoted from a lower energy level to a slightly higher level. Each such electron promotion to a higher energy level is a bit like taking an elevator. Some energy is required to raise the electron to that higher level. The important fact for our purposes is that the energy required for the electron promotion is quite specific. And, where does that specific energy come from? Quite simply, it is provided by an incoming light photon of that same energy. In Appendix A, you will find more detail as to the electronic structure of the *transition metal ions* that are the key players in producing color in glasses. The specific energy rise for an electron in a transition metal ion is provided by the absorption of a photon of that same energy. The resulting color of glass is that part of the visible light spectrum that remains once another part is stripped out by the absorption mechanism. Figure 10.11 illustrates this for a green glass, as commonly used for many wine bottles (Figure 10.12). You can compare this green spectrum with the blue spectrum created by cobalt in Figure 7.8.

Table 10.1 summarizes some of the ions that can provide the specific absorption characteristics to produce various colors in glass bottles, and Figure 10.13 shows a variety of commercial examples (including a clear bottle without color due to the absence of such coloring ions). One will notice in Table 10.1 that a given ion can produce more than one color. This variability is the result of that ion being present in different concentrations and within glasses

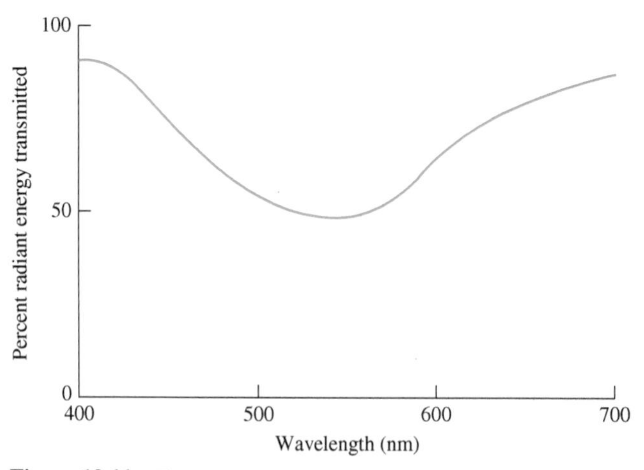

Figure 10.11 The transmission of light across the visible light spectrum in a green glass, such as that in many commercial wine bottles, is distinctly different from the spectrum of the cobalt blue glass in Figure 7.8. The absorption of light at the middle of the spectrum produces the resulting green color of the glass. A green wine bottle is shown in Figure 10.12. (Kingery *et al.*, 1976. Reproduced with permission of John Wiley & Sons.)

Figure 10.12 A green wine bottle corresponding to the absorption curve of Figure 10.11.

of different overall compositions. As pointed out in Chapter 7, the atomic-level explanation for this fact is that the specific energy levels of each transition metal ion are sensitive to the immediate arrangement of the oxygen ions surrounding it. These different arrangements follow from the different glass compositions leading to subtle changes in the energetics of the inner core electron levels and, as a result, cause that ion to absorb different portions of the visible spectrum;

Table 10.1 Colors Produced by Transition Metal Ions in Silicate Glasses

Atomic number	Ion	Color
22	Ti^{3+}	Violet-purple
23	V^{3+}	Yellow-green
24	Cr^{3+}	Green
25	Mn^{3+}	Purple
26	Fe^{2+}	Blue green
	Fe^{3+}	Pale yellow-green
27	Co^{2+}	Intense violet-blue
28	Ni^{2+}	Yellow, brown
29	Cu^{2+}	Blue, green

Source: After K. Nassau, *The Physics and Chemistry of Color: The Fifteen Causes of Color*, John Wiley & Sons, Inc., New York, 1983.

Figure 10.13 A variety of wine bottles showing the range of colors used in contemporary wine storage.

hence, producing different net colors. Another example of the importance of the ionic surroundings is that iron ions can give a strong brown color (sometimes called "beer bottle" brown) when both sulfur and oxygen ions surround the Fe^{2+} and Fe^{3+} ions, as a result of adding iron–sulfur compounds as the coloring agents.

These last four chapters have given us an overview of how modern glassware is made and how the atomic- and microscopic-scale structure helps us to understand its resulting properties (both mechanical and optical). Now, we can use this understanding to better appreciate the nature of the most ubiquitous uses of glass in the wine world – bottles and stemware.

BIBLIOGRAPHY

Callister, William and David Rethwisch, *Materials Science and Engineering: An Introduction*, 9th Edition, John Wiley & Sons, Inc., New York (2014).

Kingery, W. David, H. Kent Bowen, and Donald Uhlmann, *Introduction to Ceramics*, 2nd Edition, John Wiley & Sons, Inc., New York (1976).

Shackelford, James, *Introduction to Materials Science for Engineers*, 8th Edition, Pearson, Upper Saddle River, NJ (2015).

Smrček, Antonín, "Compositions of Industrial Glasses," in *Fiberglass and Glass Technology*, F. Wallenberger and P. Bingham, Eds., Springer, New York (2010).

Chapter 11

The Shape of Things – I. Why Bottles Look the Way They Do

"The eventual unit that is enjoyed or not, judged better or worse, is a bottle."

Hugh Johnson, *Wine: A Life Uncorked*, University of California Press, 2005. [p. 13]

The shapes of bottles are myriad and have come to symbolize the wide variety of wines that have evolved worldwide. In Figure 3.8, we saw an early example of an "English bottle" at the dawn of what we can term the modern era of wine bottles. While crudely made by today's standards, it was a shape not unlike a contemporary bottle and was cylindrical and easily stacked. The precursor to that shape was the pear-shaped bottle made in England and Germany with a long neck and wide base. The balloon-like shape was an obvious and natural one for a glassblower with minimum final shaping required. The Bocksbeutel bottle from the Franconia region of Germany was of the bulbous shape, one that still enjoys protected status by the European Union for wines from that region (Figure 11.1), although contemporary bottles are made by machine. The Bocksbeutel was, in fact, flattened somewhat so that it could be stacked and stored more easily, and this shape is still familiar in some wines from modern Portugal. (Some readers may remember the romantic Portuguese rosé, Mateus, popular in the 1970s that came in a similarly bulbous bottle.)

The bulbous-shaped bottles were not only primitive in form, but they were often fragile. Glassmaking technology was also relatively primitive, and the bottles often broke under routine handling. To appreciate the source of this fact, we

The Glass of Wine: The Science, Technology, and Art of Glassware for Transporting and Enjoying Wine, First Edition. James F. Shackelford and Penelope L. Shackelford.

Figure 11.1 The machine-made modern Bocksbeutel recalls the shape of the early bottles made by glassblowers and contains contemporary wines from the Franconia region of Germany.

can bring together concepts raised in Chapters 7, 9 and 10. In Chapter 7 on glassmaking, we saw that the decomposition of Na_2CO_3 to form Na_2O produces CO_2 gas bubbles that help to homogenize the molten glass. If the glass melt is not sufficiently fluid however, some of these gas bubbles can be trapped in the final product, and we have seen the potentially dire consequences of such bubbles in the last two chapters. Chapter 9 introduced the term "stress risers" for such defects that can contribute to the fracture of such brittle materials, and Chapter 10 showed how small concentrations of bubbles can diminish the transparency of glass.

It is the breaking of glass due to stress-rising bubbles that is our primary concern here. Through history, the problem of glass bottles breaking got to be so bad that the name of one such bottle became synonymous with a complete and ignominious failure. So much so that when Pulitzer Prize-winning author Thomas Ricks wrote a scathing criticism of the American war in Iraq in 2006, he entitled his book with the single word: **Fiasco**. This symbol of humiliation is the name of the early, bulbous Italian bottles that contained the Tuscan Chiantis.

Figure 11.2 The iconic woven straw covering for this fiasco evolved from the necessity to provide a barrier to protect the fragile glass vessels that contained the early Tuscan Chiantis.

The characteristic woven straw covering on these bottles served the simple function of cushioning these fragile vessels from impact damage (Figure 11.2). At times, both olive oil and wine were stored in these bottles (the Italian plural being *fiasci* or *fiaschi*). Some Tuscan trattorias are still known as fiascheterrias. These rustic, workingmen's taverns serve cheap and hearty food and wine. And those straw-covered Chianti bottles survived into the 1970s, joining the bulbous Mateus bottles as symbols of romance and European charm to a generation of Americans just beginning to develop a wine culture.

These simple blown glass shapes were not confined to Europe. From the sixteenth century, Catholic missionaries to South America had transported wine in earthen jugs (*Botijas*) roughly five gallons in size. By the twentieth century, these vessels had given way to glass jugs (*Damajuanes*) that had been used to import conserves of sweets and whiskey from Europe. The Damajuanes became the symbol of inexpensive bulk wine for everyday quaffing. Comparable vessels are on display today at the Quevedo family winery in the Duoro Valley of Portugal (Figure 11.3).

Figure 11.3 Large glass jugs that once held inexpensive, bulk wine make a striking display at the Quevedo family winery, Quinta da Senhora do Rosário, in the Duoro Valley of Portugal, as well providing ample storage space for wine corks. (Reproduced with kind permission of Oscar Quevedo, Quinta da Senhora do Rosário.)

The eighteenth century saw dramatic changes in the shape of wine bottles. In the early part, the neck was shortened and the sides were shaped into a more cylindrical form, as noted above. This geometry became popular not only for wine but for the storage of medicines, perfume, liquor, and cider. The eighteenth century also saw the volume of the bottles become more uniform. From the fifteenth through seventeenth century, the capacity of wine bottles varied widely between 470 ml and over 1500 ml (16–52 ounces). Sir Boyle Roche, a member of the Irish House of Commons best known for malapropisms and mixed metaphors, was unusually succinct in saying "A quart bottle should hold a quart." Bottle manufacturers seemed to heed this call and became more consistent, making bottles close in size to today's standard 750 ml that translates to 25.36 ounces in the old English measure. (A quart is about 95% of a liter, so the 750 ml bottle is exactly three-quarters of a liter and roughly three-quarters of a quart.) Figure 11.4 shows an exhibition of bottle-making, including a mold, at the Corning Museum of Glass.

Today, many wineries provide alternatives to the standard 750 ml bottle in simple multiples, usually a half bottle of 375 ml for those not wanting to deal with storing an opened bottle after a meal or a magnum of 1500 ml for serving a large group. Occasionally, magnums for special occasions are available in even larger multiples of 750 ml. An especially wide spectrum of sizes has evolved for Champagne bottles including large formats intriguingly named in the late nineteenth century after various Biblical kings. The largest of these bottles are hand-blown and rarely shipped outside of France. This grand spectrum of sizes varying overall by a factor of 160 is summarized in Table 11.1.

Figure 11.4 These bottles and bottle mold are part of an exhibit in the Innovation Center at the Corning Museum of Glass. The overall steps in the operation of the individual section (IS) bottle-making machine are summarized in Figure 3.11. (Reproduced with kind permission of the Corning Museum of Glass.)

As we noted in the first chapter, the deep indentation (or "punt") on the bottom of Champagne and other sparkling wine bottles serves a pragmatic technological purpose of providing an even distribution of the pressure developed during the second fermentation involved in the making of this type of wine. In fact, most wine bottles now have at least a minor punt facilitating handling whether or not internal pressurization is involved. These indentations originated with the glassblowers as a way to avoid the jagged pontil mark left from the glass blowing process that could scratch a table surface. The punt remained as a feature of bottle design after bottle-making became a machine process due to the stability of the upright bottle containing this indentation.

Table 11.1 Nomenclature and Volumes of Champagne Bottles

Name	Volume
Split (quarter-bottle)	187.5 ml
Half-bottle	375 ml
Bottle	750 ml
Magnum	2 bottles (1.5 l)
Jeroboam	4 bottles (3 l)
Methuselah	8 bottles (6 l)
Salmanazar	12 bottles (9 l)
Balthazar	16 bottles (12 l)
Nebuchadnezzar	20 bottles (15 l)
Melchior	24 bottles (18 l)
Melchizedek	40 bottles (30 l)

While the liquid volume content of the typical wine bottle has become largely standardized at 750 ml, the exact shape of these bottles still varies significantly. We touched on the source of this variation, a combination of tradition and pragmatics, briefly in the first chapter. While the tall, thin neck of a German Riesling bottle and the sloping shoulder of a French Burgundy bottle may be iconic shapes steeped in tradition, the square shoulder of the French Bordeaux bottle serves the pragmatic function of allowing the removal of sediment when a well-aged wine is poured slowly. The Burgundy and Bordeaux shapes have become nearly standard for their respective grapes worldwide (pinot noir and cabernet sauvignon/merlot, respectively). Reds from the Rhône area of France tend to come in bottles with shapes similar to the Burgundies. A side-by-side comparison of the three main bottle shapes is given in Figure 11.5.

Bottle shape is not the only visual clue as to the contents within. The color of bottles often follows traditions. (Recall the chemistry behind coloring glass first introduced in Chapter 7 and then in more detail in Chapter 10.) Red wines from Bordeaux typically come in dark green bottles, while the dry white wines come in light green, and sweet whites in clear bottles. For wines from Burgundy and the Rhône, dark green is common. White wines from the Mosel and Rhine areas of Germany and the Alsace region of France tend to come in bottles that are either medium to dark green or amber. Champagne is traditionally bottled in medium to dark green, except for rosé Champagnes that may come in a colorless or light green bottle.

As noted in Chapter 10, the color of glass is determined by its chemistry. With an increasingly global economy and the increasing emphasis on recycling

Figure 11.5 The three most distinctive and common wine bottle shapes are (from left) the square shoulder of Bordeaux-style wines, the sloping shoulder of Burgundy and Rhône-style wines, and the tall slender shape characteristic of Mosel (Germany) and Alsace (France) wines.

of used bottles, a relatively consistent compositional profile is now used for container glass worldwide. A typical clear glass bottle composition (recall the general "soda-lime-silica" term) is then about 13 wt% Na_2O, 10 wt% CaO, and 72 wt% SiO_2. These amounts sum to 95 wt%, with the remaining 5 wt% coming from various oxides that either come along as impurities in raw materials or are purposeful additions, especially about 2 wt% Al_2O_3 to provide chemical resistance. This clear glass composition accounts for about half of all container glass production. Amber glass bottles derive their color from Fe_2O_3 and SO_3 additions to the glass batch, with the Fe_2O_3 level at about 0.26 wt% and the SO_3 level between 0.02 and 0.05 wt%. Ironically, the presence of Al_2O_3 along with Fe_2O_3 degrades the color creating a "dirty brown" effect, so the Al_2O_3 content is reduced in these products.

Green glass bottles are produced by a synergistic combination of Fe_2O_3 and Al_2O_3, with a higher concentration of Fe_2O_3 (about 0.36 wt%). While light green bottles are colored by Fe_2O_3 alone, a rich emerald green color is produced by the additional presence of about 0.15 wt% Cr_2O_3.

The shape and color of bottles are not the only features that have become traditional symbols of wines from various regions. The mass of a bottle is another, more subtle feature. Often prestigious wines come in distinctly weighty bottles. It is not uncommon for a fine Bordeaux, Burgundy, or Rhône wine bottle to weigh as much as 1200 g (more than 2.5 lb). This hefty mass does not include the wine inside the bottle. The standard volume of 750 ml would correspond to exactly 750 g if the liquid were pure water. The small amount of solids in wine makes the beverage weigh slightly more than 750 g, but much less than the container's weight in many cases. The subliminal message is literally the *gravitas* of this important wine.

In spite of the possible marketing benefit of a heavyweight bottle, concerns over the reduction of energy consumption in glass bottle manufacturing have occupied some of the leading glass researchers such as Professor Emeritus Arun Varshneya of Alfred University in New York State. Alfred University has one of the highest concentrations of researchers devoted to ceramic and glass research in the world. Varshneya has noted that reducing the thickness of bottles by 40–50% could reduce their weight by 30–35%. Given the nearly three-fold smaller density of polymers compared with oxide glasses, the most optimistic reduction of glass bottle dimensions still leaves glass containers heavier than plastic alternatives. Nonetheless, consumer preference (*demand!*) for glass bottles for all but the most ordinary of wines makes the effort to reduce bottle dimensions a major part of the wine industry's efforts to reduce energy consumption.

Progress from this effort to reduce bottle weight faces some significant technological challenges. Many wine bottles contain not only the wine but also substantial pressures of associated CO_2 gas. The pressure within a Champagne bottle is three times that in an automobile tire. For these containers especially, reduced wall thickness needs to be accompanied by increased glass strength.

The infamous "stress risers" of Chapter 9 are an obvious focus of concern in any weight reduction program.

Champagne and other sparkling wines represent an especially tradition-bound part of the wine industry, but even major producers in the Champagne region of France have been pursuing weight reduction in the iconic and rather heavy Champagne bottle (Figure 11.6). Since the early 1970s, the standard Champagne bottle had weighed 900 g (2 lb). In 2010, members of the Comité Interprofessionnel du Vin de Champagne (CIVC), the trade organization of this celebrated region, began pursuing a "green" campaign to reduce the industry's carbon footprint. There is some energy reduction in simply manufacturing a bottle containing less mass of glass, but considerably greater savings are possible from the attendant reduction in transportation costs.

While local, sustainable agriculture has been a significant movement in recent years, the international market for both Champagne-style (sparkling) and

Figure 11.6 The Champagne bottle is both distinctive and distinguished but has also been undergoing a much needed "weight-loss plan." Wine bottle manufacturing represents an opportunity for enormous energy savings by the simple reduction of bottle wall thickness.

still wines has remained robust. At homes and restaurants where local foods are featured, those gourmet foods are often accompanied by cabernet sauvignons from Bordeaux, France, sauvignons from the Tre-Veneto of Italy, or tempranillos from Rioja, Spain. Even the consumption of "domestic wines" often involves substantial transportation costs across the continental United States; an ice wine from northern New York state may be on the dinner table in Oregon and a California cabernet in a Manhattan restaurant.

Within the Champagne region, the CIVC members started using a new 835 g bottle in 2011, and following 3 years of fermentation, it was in the hands of customers worldwide. This relatively subtle modification is estimated to reduce carbon emissions by 8000 metric tons annually (the equivalent of removing 4000 small cars from the world's highways).

Most of these new bottles are being manufactured at a St. Gobain factory in the Champagne region. The smaller quantity of glass in this new bottle reduces the carbon emissions in manufacturing by 7%. The transportation costs of the smaller bottles are also reduced. The smaller dimensions allow 2400 more bottles to be placed in delivery trucks significantly reducing the number of shipments needed per season.

This concern over the heft of bottles is worldwide. In 2010, the Tablas Creek Winery in Paso Robles, California, a pioneer in bringing Rhone varietals to North America, nearly cut the weight of their bottles by half. Their sloping shoulder Burgundy/Rhône style bottle was trimmed from 893 g (31.5 ounces) to 468 g (16.5 ounces) for both red and white wines. The winery acknowledged that the old bottles were sometimes too large for home wine racks, and they proudly announced that the 8000 cases of wine produced in the new bottles consumed 41,000 kg (90,000 lb) less glass. For the consumer that translates to each case to be carried home being 5 kg (11 lb) lighter. (Some of these bottles are seen in Figures 10.12 and 10.13.)

While we commented on large format bottles in regard to bulk wine for daily use in the distant past (Figure 11.3) and the large number of large format Champagne bottles with Biblical names (Table 11.1), large format bottles are still popular around the world for providing daily wines for local consumers. Such bottles also play a role in the training of future winemakers, as shown in Figure 11.7.

Examples of making glass bottles thinner and lighter are expanding steadily in an increasingly energy-sensitive world and are being joined by efforts to apply alternative packaging materials from other beverage industries to wine. The potential for these challenges to the dominance of glass for wine bottles will be explored further in Chapter 15.

While wine consumers expect fine and not-so-fine wines to come in glass bottles, no matter what the shape, color, or heft, they are equally adamant to drink them from a glass vessel. We now move on to explore the evolution of the shape of wine glasses, a journey as complex and ongoing as that for the bottles themselves.

Figure 11.7 (a) Large format wine bottles wait to be filled by student projects in the Teaching Winery of the Robert Mondavi Institute (RMI) for Wine and Food Science on the University of California, Davis campus. (b) Once filled, these bottles provide for numerous analyses as part of the education of future winemakers. (Reproduced with kind permission of Andrew L. Waterhouse, Robert Mondavi Institute for Wine and Food Science.)

BIBLIOGRAPHY

Alderman, Liz,"Champagne Producers Aim for 'Greener' Bottle," *New York Times*, August 31, 2010.

Bossche, Willy van den, *Antique Glass Bottles: Their History and Evolution 1500–1850*, Antique Collectors Club, Suffolk (2001).

Ellis, William S., *Glass*, Bard/Avon, New York (1998).

Johnson, Hugh, Dora Jane Janson, and David Revere McFadden, *Wine, Celebration, and Ceremony*, Cooper-Hewitt Museum, New York (1985).

MacNeil, Karen, *The Wine Bible*, 2nd Edition, Workman, New York (2015).

Pedersen, B. Martin, Ed., *Bottle Design: Beer, Wine, Spirits*, Watson-Guptill, New York (1997).

Shackelford, James, *Introduction to Materials Science for Engineers*, 8th Edition, Pearson, Upper Saddle River, NJ (2015).

Smrček, Antonín, "Compositions of Industrial Glasses," in *Fiberglass and Glass Technology*, F. Wallenberger and P. Bingham, Eds., Springer, New York (2010).

Taber, George M., *To Cork or Not to Cork: Tradition, Science, and the Battle for the Wine Bottle*, Scribner, New York (2007).

Chapter 12

The Shape of Things – II. The Rise (and Fall?) of Varietal-Specific Stemware

"Before there was a real estate bubble or the cult wine bubble, I had a wineglass bubble."

James Laube, *Wine Spectator*, September 30, 2009

The English language is both complex and wonderful but, at times, a bit maddening too. As we begin this chapter on the glass vessels from which we enjoy the wines of the world, we face the interesting dilemma that the vessels and the material from which they are crafted are described with the same word, *glass*. In fact, we lower our *glasses* from our forehead to peer at the label of a wine bottle before we pour it into our *glasses* and reflect on how these elegant vessels were fabricated from this mysterious, amorphous material: *glass!*

In any case, a discussion of glass vessels for drinking wine soon turns to one name: *Riedel*, the eponymous name of the most famous wine glass manufacturer with a legacy of over 250 years. Today, they market themselves as "The Wine Glass Company." The history of the Riedel family reads much like a Dostoyevsky novel, beginning with Johann Christoff Riedel whose 1729 robbery and murder (the result of the glass trader traveling with a bit too much money on his person) may have inspired another work of fiction, Friedrich Schiller's 1797 novel *The Cranes of Ibykus*. The location of the Riedel factory would change from its original Northern Bohemia location to its current one in Kufstein, Austria (Figure 12.1) as the family fortune rose and fell over the years against a backdrop of war and peace across Europe.

The Glass of Wine: The Science, Technology, and Art of Glassware for Transporting and Enjoying Wine, First Edition. James F. Shackelford and Penelope L. Shackelford.

Figure 12.1 The headquarters of Riedel along with its glass factory are in Kufstein, Austria. (Reproduced with kind permission of Riedel Glas Austria.)

The oscillations of the Riedel family fortunes reached a maximum of sorts in the early part of the twentieth century as Walter Riedel, the eighth generation of the family and a major European industrialist, was pressed into service by the Ministry of Aviation of the Third Reich during World War II to produce state-of-the-art fiberglass. His technical prowess was exceptional, but his misfortune of being on the losing side led to his exile after the end of the war to Eastern Siberia beginning a 10-year ordeal that included helping to rebuild Russian glass factories. His personal misfortune was compounded as the Riedel family fortune had been seized in 1945 by the newly formed state of Czechoslovakia. While Walter was incarcerated in the Soviet Union, Claus Josef Riedel of the ninth generation was able to resettle in Austria. While essentially penniless, he nonetheless was able to enlist the support of another glass dynasty, Swarovski, which had decided not to take over a bankrupt stemware factory in Kufstein near Innsbruck. The head of the Swarovski family had been mentored by Claus's great grandfather Josef, and he took young Claus in, financed his studies in chemistry, and ultimately advanced the young Riedel funds to revitalize the Kufstein facility.

Upon his release from Soviet imprisonment, Walter joined his son in the Kufstein operation. As a practical matter, the two differed in emphasis, with the younger Riedel preferring to focus on wine stemware. It was this ninth generation Riedel, Claus, who prevailed and used this platform to launch a revolutionary new style of glassware for wine drinking that would become the world standard: a simple, unadorned, and utilitarian form very much influenced by the Bauhaus tradition. In a fitting tribute to his contribution, the Museum of Modern Art in New York has a set of his elegantly simple 1958 glasses as part of its permanent collection.

Claus's aesthetic creativity produced what is now the iconic symbol of modern wine stemware design, but his scientific training would play an equally important role in a systematic set of experiments carried out with a set of

Figure 12.2 (a) Claus Josef Riedel, ninth generation member of the eponymous wine glass company displaying his simple stemware design that has become the standard for contemporary wine appreciation. (b) Claus Josef (ninth generation, left) and Georg Josef (10th generation, right) are seen toasting eighth generation Walter Riedel. (Reproduced with kind permission of Riedel Glas Austria.)

sommeliers and glassblowers to demonstrate the concept that the bouquet, taste, balance, and finish of wines are affected by the specific shape of the wine glass. This novel concept was launched in a new product line in 1961, and over the years, Riedel has developed over 25 varietal-specific shapes not counting additional forms for water and various spirits. The hand-made Sommelier Series was launched in 1973, and his son, 10th generation Georg Josef, took the helm of the company in 1987 bringing a greater business expertise and mechanizing the production of stemware with the introduction of the considerably more affordable Vinum Series. These three generations of Riedels are shown in Figure 12.2, and Figure 12.3 shows a range of stemware shapes from the Vinum line. With the encouragement of Robert Mondavi, Georg launched a Riedel company in the United States now run by his son, 11th generation, Maximilian. This youngest

Figure 12.3 Distinctive shapes for different red and white wines are but a few of the more than 25 varietal-specific wine glasses developed by the Riedel glass company. These glasses are from the machine-made Vinum Series. (Reproduced with kind permission of Riedel Glas Austria.)

Riedel, with little room for innovation in stemware design, given the current stable of over 25 shapes has launched a new, radical concept – the stem-less wine glass, a standard tapered bowl but with a flat bottom without the usual stem and base. The elimination of the fragile stem has been welcomed, but the non-traditional feel of the glass and the tendency to warm the wine in the hand as well as some awkwardness while washing it has produced a mixed reaction.

While even serious wine connoisseurs might be hesitant to maintain a stable of 25 different shaped glasses, one by-product of Claus Riedel's research was the nearly universal maintaining of a double set of stemware in the home and at restaurants: large glasses for red wines and smaller ones for whites. Today, we might feel this epicurean rule to be a canonical law given its commonplace practice. In fact, the "rule" is more a pragmatic average of the results obtained by Claus Riedel and his band of sommeliers and glassblowers as most red varietals performed best in larger glasses *on average* than did the whites. Thoughtful observers, however, often noticed that highly aromatic whites benefited as well from larger bowl stemware, and that reds without a powerful "nose" were not particularly handicapped by being served in a smaller glass. In the last decade or so, such observations have led to a widespread recommendation among a range of prominent wine journalists remarkably similar in its simplicity, namely, serve all wine in a single, ample-sized (about 600 ml or 20 ounce) glass with tapered bowl of the basic shape pioneered by Claus Riedel in the late 1950s. This egalitarian group further suggests that there is little advantage in the more expensive and delicate hand-blown glasses with machine-produced stemware serving quite adequately, certainly for everyday enjoyment. James Laube of the *Wine Spectator*, Karen MacNeil author of the *Wine Bible*, the husband-and-wife team of former wine critics for the *Wall Street Journal*, Dorothy Gaiter and John Brecher, Eric Asimov of the *New York Times* and the eminent British wine writer Jancis Robinson, are all part of this chorus. Figure 12.4 is an example of

Figure 12.4 The people's choice? The authors offer a toast at the Taco Maria Restaurant in Costa Mesa, California, with a moderately priced machine-made wine glass, of the type that a number of wine critics have found to be adequate for general wine drinking (red and white). (Reproduced with kind permission of Daniela Wood.)

this one-size-fits-all approach. Some are even caustic in their comments. Andrew Barr, author of *Wine Snobbery*, calls the voluminous Burgundy goblet "a joke." James Laube admits sheepishly that, at one time, he had a pair of wine glasses (one each in the Bordeaux and Burgundy shapes) that could hold the entire contents of a 750 ml bottle. They proved to be so large and clumsy (and in his word "obnoxious") that he never used them.

While the bulbous shape of the wine glass in Figure 12.4 is extremely common in homes and restaurants worldwide, Italian designers and glassmakers have tended to present a slightly modified shape in which some angularity is included in the bowl, as illustrated in Figure 12.5.

The many variations in wine glass shape is no small problem for those engaged in sensory evaluation, as we shall see shortly in reviewing the limited

Figure 12.5 Italian glassmakers typically produce stemware with a characteristic angularity in the bowl designs, distinctive from the purely bulbous shape in Figure 12.4. (Reproduced with kind permission of Kim Goodwin, Luigi Bormioli Italy.)

Table 12.1 Dimensional Comparison of the ISO 3591 "Standard" Glass[a] and a Typical Commercial Stemware Glass[b]

Average dimension	ISO 3591	Commercial stemware
Bowl opening diameter	46 mm	75 mm
Bowl diameter (maximum)	65 mm	96 mm
Maximum diameter/opening diameter ratio	1.4	1.3
Bowl height	100 mm	140 mm
Overall height	155 mm	226 mm
Bowl height/overall Height ratio	0.65	0.62
Base diameter	65 mm	86 mm

[a]Dimensions provided by Bormioli Luigi, Parma, Italy.

[b]Riedel Cabernet Sauvignon/Merlot (Bordeaux) Glass, Vinum Collection.

literature on the effect of glass shape on wine appreciation. In an attempt to eliminate the wine glass geometry as a variable in wine tasting, the **International Standardization Organization (ISO)** has developed "standard" glass, the egg-shaped ISO 3591. While the general appearance of the ISO 3591 is quite similar to commercial stemware, Table 12.1 compares the dimensions of the ISO 3591 with a "one-size-fits-all" glass of the type shown in Figure 12.4. One will notice that, while the standard glass is smaller, the overall proportions ([maximum bowl diameter/bowl opening diameter] ratio and [bowl height/total height] ratio) are quite similar. We will see later that the maximum bowl diameter/bowl opening diameter ratio is one shape metric that seems to have a measurable effect on the wine tasting experience.

Although the debate over the size and shape of stemware for still wines is ongoing, there is a wide (but not quite unanimous) acceptance of the characteristic narrow dimension of the Champagne flute as appropriate for Champagne and similar sparkling wines. In fact, the fluid mechanics of bubble formation along the wall of these small vessels is well understood and supports the appropriateness of avoiding a large bowl. Nonetheless, Karen MacNeil quotes Remi Krug of Krug Champagne as preferring the slighter wider tulip-shaped white wine glass that provides stronger aromatics at the expense of bubble formation. (He is satisfied with the bubbles "dancing in the mouth.") In any case, a practical aspect of this bubble formation is that one should *not* swirl Champagne as part of the tasting ritual. Swirling simply accelerates the loss of its distinctive effervescence. Similarly, the saucer-shaped Champagne coupe often used for wedding toasts is a disastrous choice, allowing bubbles to dissipate quickly as well as facilitating the rapid warming of the shallow bowl contents by the drinker's hand. Both the traditional flute and the tulip-shaped glass are shown in Figure 12.6.

Another specific glass design for a particular wine is the official Port glass shown in Figure 12.7. In this case, the glass represents a significant

Figure 12.6 The traditional flute and the slightly wider, tulip-shaped glass are ideal for Champagne and similar sparkling wines.

aesthetic choice by the Port Wine Institute of Porto, Portugal, with the design commissioned to the eminent Portuguese architect Alvaro Siza who gave the small glass (appropriately so for the fortified wine with a typical alcohol content of 20 vol. %) a distinctive square cross-section stem with an indentation for the thumb.

While Riedel and other manufacturers of stemware have marketed the benefits of varietal-specific glasses for many years, various criticisms have been voiced at times, beyond the suggestion noted above that a one-size-fits-all approach is adequate if not preferable. Daniel Zwerdling summarized the critiques in an article entitled "Shattered Myths" published in *Gourmet* magazine in 2004. Zwerdling focused on 10th generation Riedel patriarch Georg Riedel, praising him as a master communicator and salesman who trained his sales staff to provide highly effective seminars in which customers are guided through comparative tastings of various wines in a spectrum of the Riedel varietal-specific stemware. Zwerdling's article recounted how some of the luminaries of wine appreciation, including the most eminent of wine critics Robert Parker and

Figure 12.7 The Portuguese architect Alvaro Siza designed the official glass for Port, the popular fortified wine of Portugal. Here, it is seen in a Duoro Valley tasting room (note the distinctive square cross-section stem with a thumb indentation near the top of the stem). (Reproduced with kind permission of Bernado Carvalho, Quinta Sta Eufêmia.)

Wine Spectator Executive Editor Thomas Matthews, have been convinced of the "Riedel effect" and have then provided ringing endorsements. Zwerdling, however, went on to suggest that, while Riedel's demonstrations are effective case studies in marketing psychology, they lack a scientific foundation. In this latter critique, he quotes Linda Bartoshuk of the Yale University School of Medicine as saying that a central tenet of the Riedel seminar, a "tongue map" that identifies different regions of the tongue that sense different tastes (sweet tasted at the front tip of the tongue, salt tasted on narrow strips along the side, etc.) is a concept discounted 30 years earlier. Bartoshuk, who does research on how people taste, emphasized that all taste buds are created equal; each can sense salty, sweet, sour, or bitter no matter where located on the tongue. Zwerdling reported that he had a subsequent conversation with Georg Riedel himself who acknowledged that the tongue map is not scientifically valid but nonetheless finds that it provides an easier way to explain how the glass "works" to the general public.

In his article, Zwerdling cites the research of Marcy Levin Pelchat, a scientist at the Monell Chemical Senses Center in Philadelphia, PA. Pelchat and her colleagues at Monell had been given a demonstration of Riedel glasses by Georg himself, and while impressed, Pelchat wanted to do a controlled study of her own. She subsequently did so with Jeannine Delwiche of the Ohio State University. The research of Pelchat and Delwiche was published in the *Journal of Sensory Studies* in 2002 and involved a modest set of 30 subjects from the Monell staff. The authors attempted to eliminate bias by confining the sensory experience to smell, with all 30 subjects (21 females and 9 males) being inexperienced wine drinkers who were blindfolded and constrained by a headrest that standardized the distance from the sensor and the wine. Even the swirling of the glass was mechanized by a vortex device set at a standard low setting. The study included four glass shapes: a square one, a small "standard restaurant" bulb-shaped glass, a slightly larger chardonnay-specific bulb-shaped glass, and a still larger Bordeaux (cabernet sauvignon)-specific bulb-shaped glass. The authors found no significant difference in results for a California cabernet sauvignon sampled in the four different glass shapes, except that the total intensity of the wine aroma was *lowest* in the Bordeaux-specific glass. Of course smelling wine is not a complete tasting experience, but the neutral and even negative results for comparing glass shapes in a carefully controlled experiment were not encouraging for the argument that shape should matter.

Zwerdling went on to describe how Georg Riedel nonetheless continued in his search for scientific validation by approaching Thomas Hummel, an expert on sensory analysis at the University of Dresden, and promised funding along with scientific freedom in a larger study of stemware shapes. Hummel's results were published in the journal *Appetite* in 2003. Hummel carefully selected approximately 180 subjects with approximately 90 receiving red wine and approximately 90 receiving white. As with the Pelchat and Delwiche study, untrained wine drinkers were used. Hummel used three types of glasses, all with common wine glass stem and base configurations, but only one had the classic bulbous shape popularized by Claus Riedel. The distinctly different bowl shapes were called a "beaker" shape that had an essentially fixed diameter comparable to a water glass and a "tulip" shape with a relatively small bowl diameter and an outward flared lip. Hummel honored the common practice of a larger glass for reds and a smaller one for whites by using the Riedel Chianti glass for the red wine sample and the Riedel chardonnay glass for the white. (The wines used were in fact from the Robert Mondavi winery, a cabernet sauvignon, not a Chianti, and a chardonnay.) Not being rigorous in matching the red wine sample to the specific Riedel glass seemed of little importance, as the study found no significant difference between glasses for "sweet," "salty," or "bitter" ratings, although "sourness" was found to be significantly higher in the beaker glass (recall that is the one roughly equivalent to a water glass). On the other hand, the bulbous Riedel glasses did display a higher intensity of wine odors and subsequent "increased liking of the wine odors." The reason for that latter observation was not clear but was suggested to be related to the larger volume of liquid in those glasses that could allow "trapping of odors." In the closing discussion, the authors commented ". . . it remains unclear why different glasses should modulate gustatory sensations."

A statistically significant relationship between wine glass shape and sensory evaluation was found in a Canadian study but without any varietal-specific implications. Margaret Cliff of the Pacific Agri-Feed Research Centre studied the aroma and color intensity evaluations of 18 blindfolded subjects. Unlike the other studies described previously in which the subjects lacked a significant knowledge of wine, Cliff chose a set of enology and viticulture students from Brock University in Ontario, Canada. They were blindfolded to eliminate any cognitive or visual clues and were not even allowed to touch the glasses. The study involved three different glasses, namely, an ISO 3591 (see Table 12.1), a Riedel chardonnay glass, and a Riedel Burgundy glass. The wines evaluated were a white (riesling) and a red (cabernet franc) produced at a Canadian research institute. Both "sound" and "defective" wines were used, with the defective produced by adding 2 ppm ethyl acetate. Cliff found that the aroma intensities correlated directly with the maximum diameter/opening diameter ratios for the bowls (highest in the Burgundy glass and lowest in the chardonnay). In other words, the more bulbous the glass, the more aromatic the experience; a result consistent with Hummel's and again implying that the greater volume contained within a given opening diameter will provide more aromatics. These results were *independent* of the type of wine evaluated (red, white, sound, or defective). Also, color intensities correlated with gross dimensions: maximum diameter, height, and volume. Cliff's final conclusion was that the intermediate-sized and independently designed ISO glass is best for overall, objective evaluations of wine characteristics.

While the high quality of the Riedel stemware (and stem-less ware) is widely acknowledged, the claims of benefits from subtle and not-so-subtle variations in shape for different varietal wines have yet to be confirmed in the few attempts at scientific validation. The recommendations of some prominent wine journalists, as noted above, for a one-glass-size-fits-all approach to wine appreciation are certainly a simple alternative. Wine consumers remain the ultimate judges until additional sensory studies might provide more clarity into this highly subjective topic.

And so the size and shape of wine glasses are as varied as those of wine bottles, and, in both cases, form has followed function – to some extent. Much of the variety can be ascribed to pragmatics (collecting sediment in a bottle; maximizing the aromatics from a glass) but some simply follow from tradition, aesthetic preference and, in the case of wine glasses, perceived differences in wine taste to which not all agree. Fortunately, in this case, you are the final arbiter of the shape of things.

BIBLIOGRAPHY

Barr, Andrew, *Wine Snobbery*, Faber & Faber, London (1988).

Cliff, Margaret, "Influence of Wine Glass Shape on Perceived Aroma and Color Intensity in Wines," *Journal of Wine Research*, **12**, 39–46 (2001).

Delwiche, Jeannine and Marcy Pelchat, "Influence of Glass Shape on Wine Aroma," *Journal of Sensory Studies*, **17**, 19–28 (2002).

Glanville, Philippa and Young, Hilary, Eds., *Elegant Eating: Four Hundred Years of Dining in Style*, Harry N. Abrams, New York (2002).

Hummel, Thomas, Jeannine Delwiche, C. Schmidt, and K.-B. Huttenbrink, "Effects of the Form of Glasses on the Perception of Wine Flavors: A Study in Untrained Subjects," *Appetite*, **41**, 197–202, (2003).

Johnson, Hugh, Dora Jane Janson, and David Revere McFadden, *Wine, Celebration, and Ceremony*, Cooper-Hewitt Museum, New York (1985).

Lloyd, Ward, *A Wine Lover's Glasses: The A.C. Hubbard, Jr. Collection of English Drinking Glasses and Bottles*, Richard Dennis, Shepton Beauchamp, England (2000).

MacNeil, Karen, *The Wine Bible*, 2nd Edition, Workman, New York (2015).

Shackelford, James, *Introduction to Materials Science for Engineers*, 8th Edition, Pearson, Upper Saddle River, NJ (2015).

Zwerdling, Daniel,"Shattered Myths," *Gourmet*, August, 2004.

Chapter 13

The Controversy over Cork – Glass Stoppers to the Rescue?

"Drawing a cork is like attendance at a concert or at a play that one knows well, when there is all the uncertainty of no two performances ever being quite the same."

George Asher, *On Wine* (Random House, New York, 1982)

Although this book focuses on the widespread use of glass for storing and consuming wine, another humble material has played an absolutely critical role in wine history for over 400 years: a small wooden cylinder, the *wine cork!* Figure 13.1 shows this familiar object.

The main source of **cork** for wine bottle closures is the Iberian Peninsula, especially Portugal that provides roughly half of the world supply. A typical cork tree, as shown in Figure 13.2, is, in fact, part of the oak family that was central to our discussions of winemaking in Chapter 4. In this case, it is the *Quercus suber* species that provides a thick, easily stripped outer bark. The cork crop is an example of sustainable agriculture, in that the outer bark is stripped periodically, and there is no need to cut down the tree. The bark is sufficiently thick for stripping once the tree is about 25 years old, and a regular cycle of stripping can continue every 9 years. The trees continue to grow, and the quality of the cork is higher after the first couple of harvests. A tree can live for as much as 300 years. After the bark is harvested, the cork cylinders are punched out of the bark in a direction parallel to the axis of the tree.

Cork was first used successfully as wine bottles' stopper in England in the late 1500s. Before that time, wine storage and shipping were largely done in pottery and wooden containers. We were introduced to some of the ceramic amphorae in Chapter 2. By 1500 BC, sophisticated Egyptian winemakers were sealing cylindrical tops of the amphorae with chopped up organic matter (leaves

The Glass of Wine: The Science, Technology, and Art of Glassware for Transporting and Enjoying Wine, First Edition. James F. Shackelford and Penelope L. Shackelford.

Figure 13.1 A familiar cylindrical piece of cork is a critical barrier between wine and a hostile environment.

Figure 13.2 The *Quercus suber*, or cork oak, tree provides the raw material for wine corks from its thick, outer bark. Here we see the harvesting of that bark. (Reproduced with kind permission of APCOR, the Portuguese Cork Association.)

and clay) topped off with a slab of pottery and wet clay. Three millennia later, the cork/bottle system was well established by 1703 when the signing of the Methuen Treaty between Portugal and England made this closure system an even more widespread format. As noted in Chapter 3, this treaty significantly enhanced the availability of cork for bottle sealing, but the 1795 English patent for a corkscrew was the final, pragmatic step that made the bottle/cork sealing system the standard for fine wine appreciation.

There is some irony in the fact that cork also provided the most basic term in the field of biology. Robert Hooke, one of the great scientists of the seventeenth century, was among the first to use the optical microscope to reveal structure of the material world smaller than the unaided eye can see (Figure 13.3). While viewing a cross-sectional slice of cork under a microscope, he noted the structure reminded him of the collection of small rooms in which monks lived, hence the term *cell.* The cellular structure of cork contributes to the highly desirable nature of this particular wood to compress slightly under pressure and provide the tight seal with the bottle opening.

George Taber provided the world an exclusive firsthand account of the 1976 French versus California wine tasting in his book *The Judgment of Paris*. This fascinating contest in which the upstart New World wines were the surprise winners was a major step in creating the modern global wine economy. It is interesting that his next book focused entirely on the humble wine cork and its alternatives, *To Cork or Not to Cork*. But, why "alternatives?"

Figure 13.3 What Robert Hooke saw: The fine cellular structure of cork that gives rise to the unique properties of this bottle sealing material also gave rise to the term cell when the great seventeenth-century scientist Robert Hooke, who was among the early users of the optical microscope, found the individual cells in the cork structure looked like the small rooms of contemporary monks. (Reproduced with kind permission of Fred Hayes, Advanced Materials Characterization and Testing Laboratory, University of California, Davis.)

An early challenge for cork stoppers in wine bottles was to maintain a tight seal, but, as bottle-making technology improved and bottle openings became reliably consistent, that problem largely went away. In the rare instances when this breakdown in quality control still occurs, the resulting oxidation of the wine can completely destroy its intended character. The next chapter will look more closely at the complex relationship between wine and air and the role that glass can play in optimizing that relationship.

While the destruction of wine by inadequate sealing of the wine cork is now a rare occurrence, a chemical issue remains that at times has haunted the entire wine industry. Taber's book chronicles this chemical taint in great detail. The culprit is the compound TCA (2,4,6-trichloroanisole) that in even small amounts can produce objectionable tastes and smells, leading to the term **cork taint** or simply that the wine is *corked*. The occasional presence of TCA "failures" was a traditional fact of life for wine drinkers and considered an acceptable price to pay for this generally reliable closure system. In the 1980s, however, a relaxation of the production standards within the Portuguese cork industry began to produce some disastrous and high-profile results for winemakers around the world. Among the most dramatic, the David Bruce Winery in Northern California had massive losses in its chardonnay business in 1987. A growing number of similar accounts led to accelerated efforts within the wine industry to produce a more "modern" closure system using more contemporary materials.

Much of the effort to replace the cork concentrated on the use of the metal screw cap, a technology well established for other segments of the beverage industry with an example shown in Figure 13.4. In spite of the loss of the romance associated with the ritualistic removal of the cork, the generally more reliable screw cap system caught on in the 1990s, especially with inexpensive wines and most especially with white wines. The aluminum screw cap is now widely used in the Australian and New Zealand wine industries. A considerable debate has emerged within the overall community of fine winemakers and consumers as to how effective the screw cap will be for the long-term storage of red wines. Of course, the image of opening a first-growth Bordeaux in a Michelin-starred restaurant with the simple twist of the wrist is nearly impossible to imagine, but perhaps even more important is the possibly subtle difference between the wine/atmosphere barrier provided by cork and the metallic screw cap. The absence of cork ensures an absence of TCA, but that absence may also eliminate a very small degree of oxygen permeation over decades of storage that can be essential to the nature of the wine's aging. We will return to this issue in more detail later in this chapter, but the issue of oxygen transport through the stopper requires us to acknowledge that the screw cap is not strictly metallic. A critical component within the aluminum capsule is an inner seal, with the most common one being a combination of a thin layer of tin (again, a metal) and saran. Yes, the saran in the tin/saran seal is the same saran familiar since the 1950s as Saran Wrap found in many kitchens. The original Saran™ products were developed by the Dow Chemical Company with key properties being virtually no oxygen

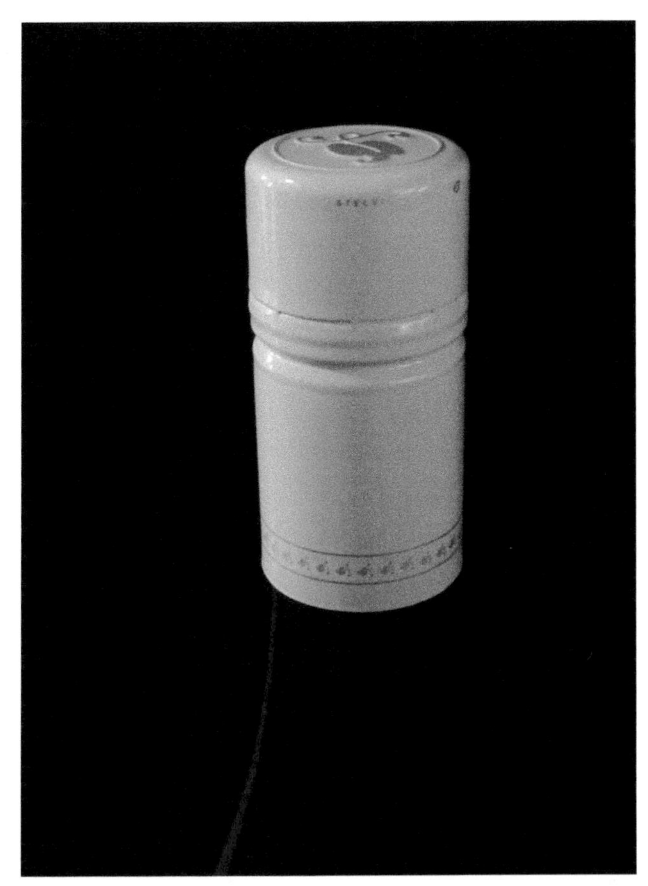

Figure 13.4 The aluminum metal screw cap closure has been since the 1990s a popular alternative to the use of cork.

or water transport. The Dow composition is polyvinylidine chloride (PVDC), a polymer still used by Dow in a wide variety of barrier materials. Ironically after Dow sold the Saran Wrap product line to the S.C. Johnson Company, the composition was changed to polyethylene, a more easily manufactured polymer and one free of chlorine, a source of some environmental concerns.

Another route to avoiding cork taint that did not displace the centuries-old corkscrew ritual was the use of synthetic cork. Various synthetic polymers and polymer/cork composites have been developed and have found reasonably wide acceptance. While these materials also can ensure an absence of TCA, they may also have a somewhat different relationship with the surrounding atmosphere than true cork. Some of the materials also have frustratingly different relationships with the corkscrew. A hard, rubbery "cork" can be notoriously difficult to

Figure 13.5 A cork for sealing a bottle of sparkling wine such as Champagne has a distinctive shape based on structural differences (an agglomeration of bonded cork chunks in the top and three bonded natural cork disks on the flared bottom) and mechanical forces (the bulbous top was not inserted into the bottle and the natural cork disks were highly compressed inside the high-pressure bottle and continue to expand once uncorked).

remove and sometimes nearly impossible to reinsert into the partially consumed bottle.

Before moving on to a glass alternative to cork, we should acknowledge that not all cork stoppers that we see are in the shape of simple cylinders. Figure 13.5 shows a more complex shape familiar to anyone who has opened a sparkling wine such as Champagne. This overall shape is sometimes compared with a mushroom. As a practical matter, this cork starts out as a straight cylinder, although of substantially larger diameter than the bottle opening. So, a 31-mm diameter cork is compressed down to the 18-mm diameter opening and partially inserted into the bottle. The bulbous top of the cork in Figure 13.5 is then the portion that is not inserted into the bottle. Close inspection of the figure shows that the top portion of the cork is composed of small chunks that have been bonded together by Food and Drug Administration (FDA)-approved binders. The bulbous top had been constrained after the cork insertion by a metal wire cage called a *muselet* to avoid any premature expulsion of the cork by the high-pressure contents. Again, inspection of the figure shows that the bottom of the cork that flares out is composed of three natural cork disks. This flaring is the result of the inserted portion of the stopper being highly compressed by the interior pressure, forming a constant pressure against the neck of the glass bottle. Once pulled from the bottle, the natural cork disks continue to expand giving the mushroom appearance.

While on the subject of sparkling wines, one can recall from the discussion of producing sparkling wines in Chapter 4 that the bottles are capped at the onset

Figure 13.6 At the Costadila Winery in the Veneto region of Northern Italy, the beer-like metal cap used in the intermediate stage of making French Champagne serves as a final closure for this Prosecco made in a traditional way.

of the second fermentation, with the cap removed for the purpose of *degorgement*, the removal of the yeasty sludge prior to the final closure with the flared-bottom cork shown in Figure 13.5. Figure 13.6 shows one of the metal cap enclosures that is reminiscent of beer bottles and is sometimes called the "crown cap." In fact, this particular metal cap is used as a final closure instead of cork in the Prosecco wine produced by the Costadila Winery in the Veneto region of Northern Italy. The Costadila wines are produced in the spirit of the traditional or "natural" wine movement discussed in Chapter 4, and using the simple metal cap is consistent with that philosophy.

While most of the attention for avoiding cork taint has focused on screw cap and synthetic cork closures, a glass system is gaining a considerable following also. A glass-to-glass seal is perhaps the ultimate solution from the perspective of this book that extols the virtues of glass as an ideal medium for wine storage. Those who have spent some time in chemical laboratories may recall the occasional use of ground glass stoppers for glass bottles. These closures, sometimes called stopples, provide very tight seals while allowing repeated openings and closings. In the absence of a thin lubricating film on the glass-on-glass interface, however, the glass stopper can "bind" and prove frustratingly difficult to remove. George Taber in *To Cork or Not to Cork* retells stories that, in fact, early winemakers did, indeed, turn to this optimal system. Taber points out that, as late as 1825, glass stopples were still the preferred luxury closures for Château Lafite, ground with emery paper to precisely fit specific bottles and tied down with string connected to a button on the neck. Unfortunately, removing these stopples frequently broke the bottle rather than merely opening it. So, the glass challenge to cork languished for nearly two centuries, until a German homeopathic doctor named Karl Matheis revisited the concept with a determination that finally produced positive results that were not limited to the luxury wine market. Matheis is an intriguing figure who had already attained financial independence with the invention of the worker's safety shoe. His inventive attention turned to wine closures when a neighbor of his, Hans Marx of Weingut Marx, told him of substantial losses of his 2001 riesling and sylvaner production due to cork taint. Dr. Matheis recalled his neighbor's woes while noticing an airtight glass ampoule used for blood transfusions that used a glass stopper topped with an aluminum cap. Within a year, the prolific doctor filed 20 patents for the concept of a glass stopper for wine bottles. Dr. Matheis was not only creative but also pragmatic in realizing that a substantial amount of research and development would be required to perfect the large-scale commercialization of the demanding task of securely sealing a wine bottle for an extended period of time. He approached Alcoa Germany that, in addition to being a subsidiary of the American aluminum giant Alcoa, was among the world's largest producers of screw cap closures. Although originally hesitant, Alcoa Germany eventually agreed to pursue Matheis' concept and invested 6 months in refining it leading to an agreement with the inventive doctor. The March 2003 ProWein show introduced the new product as Vino-Lok, and, in short order, fifty winemakers across Germany, Austria, France, and Italy agreed to evaluate the new closure. Following considerable experimentation, Schloss Vollrads ordered 1 million tall green bottles designed to especially demanding specifications. To make the closure successful, the first 5 mm of the bottleneck had to be consistently straight (Figure 13.7). If at all concave, the stopper would tend to pop out of the bottle. If convex, the stopper would be difficult to extract. By January 2005, Schloss Vollrads bottled half of a tank of riesling with cork and the other half with Vino-Lok. The public response was impressive. The sales in February were 80% Vino-Lok. By March, 90% of sales were Vino-Lok, and by May they were 95%. Sensing an attractive product, Alcoa developed full-scale, automated production of Vino-Lok in their Worms, Germany plant. In contrast to the business model generally followed by the users of screw caps that focused on inexpensive

Figure 13.7 This long-necked bottle of a German riesling is closed not by cork but by a sophisticated and precisely sized glass stopper.

wines, the Vino-Lok was promoted for relatively expensive wines. Each closure costs about 50 cents, similar to that for premium cork.

But why, at long last, were these glass closures successful and the laboratory-like ground glass stoppers used in the early nineteenth century disastrously impractical? Not unlike the dependence of screw caps on a small amount of polymer (PVDC), the secret of the modern glass stopper's success lies in the fact that Vino-Lok is not entirely glass. A polymeric o-ring is seated in a groove just under the top of the stopper, as shown in Figure 13.8. Alcoa's experience in food packaging proved helpful in this regard, leading to the choice of a DuPont ethylene vinyl acetate polymer with the brand name Elvax. The o-ring eliminates the need for intimate contact between glass surfaces and subsequent seizure (alas, those broken Château Lafite bottles!). Elvax appears to be an ideal material selection. The halogen- and plasticizer-free polymer is highly elastic, durable, insoluble in alcohol, resistant to other wine constituents such as weak acids, and provides no apparent aftertaste.

Whitehall Lane winery in the heart of the Napa Valley was an early, major user of Vino-Lok glass stoppers in the California wine industry. Because a screw cap closure used in the region had the brand name Vinloc, the Vino-Lok name was changed to Vino-Seal. The American version also used a traditional foil capsule over the closure as opposed to a less elegant aluminum overcap used in

Figure 13.8 A critical component of the glass stopper design is the polymeric o-ring at the top of the tapered section and below the wide cap.

the original European version. The enthusiasm for the new stopper at Whitehall Lane led, in April 2006, to the announcement that their 2003 Reserve Cabernet (valued at $75) and their Leonardini Vineyard Cabernet ($100) would be released with Vino-Seal closures. Apparently, Whitehall Lane has received no complaints about this significant departure from the traditional use of premium cork for fine wines. Some other winemakers are also viewing screw caps and glass stoppers as an ideal solution for the sealing wine bottles. On the other hand, the venerable Ridge Winery in California did report that the use of a glass stopper resulted in a "lack of oxygen" interacting ever so slightly with the wine keeping it from being as mature and complex eventually as the counterpart aged in natural cork. Ridge is not alone, and so the issue raised with the use of the metallic screw cap is seen again with glass. Wine appreciation is highly subjective, and the debate over closures and the amount of oxygen that might find its way to the wine only adds to the subjectivity.

These subjective observations of a lack of oxygen from both screw caps and glass stoppers have led to a number of scientific studies of various closure systems. Two 5-year studies are worth noting, although we must also acknowledge that these concerns over long-term effects are for age-worthy wines. Most wines are thought to be consumed within 24 h of purchase, but, for prized wines, aging after purchase can involve years and even decades. So, while the history of alternative closures is relatively recent, 5-year comparisons have been completed that give some clues as to the nature of closure effects. As we review these studies, we can recall the discussions of aging raised in Chapter 4 on winemaking, in which we saw two broad issues – (a) biochemical changes occurring within the wine and (b) the additional effect of *oxidation* that can affect the nature of those long-term biochemical changes in a subtle and perhaps positive way or in a not-so-subtle and destructive way.

A study from the Australian Wine Institute published in 2005 followed two whites, a riesling and a chardonnay exposed to oak, after 5 years in a variety of closures: screw cap, natural cork, synthetic cork, and glass ampoule (an oxygen-free sealed vial as used in the pharmaceutical industry, comparable to Dr. Matheis' original inspiration for the glass stopper). Consistent with the growing literature on this subject, the screw cap (and glass ampoule) showed the least oxidation, and the synthetic cork showed the most. On the other hand, the oxygen-starved wines under the screw cap and within the glass ampoule both demonstrated a "struck flint/rubber" aroma. In contrast, the more oxygen permeable synthetic cork demonstrated a characteristic oxidized aroma along with noticeable darkening (a brown color) and a reduced SO_2 level in the wines. These effects were seen within the first 3 years of the study.

A 5-year study from UC Davis published in 2015 under the supervision of Professor Andrew Waterhouse monitored the effect of the oxygen transfer rate (OTR) through synthetic cork on a red wine, cabernet sauvignon. The study used two types of synthetic closure materials from a major supplier to the wine industry, Nomacorc. The two closures represented the extreme range of OTR values from the supplier: Nomacorc Light with an OTR of 16 μg/day and Nomacorc Premium with an OTR of 5 μg/day, a factor of more than three times. Although confined to a synthetic closure, the wide range of oxygen permeability has implications for various closure systems. In any case, the central finding of the study was that the higher OTR closure corresponded to significantly greater loss of SO_2, which the researchers emphasized is "the first line of defense towards oxidation." A by-product of the reduced SO_2 levels was greater color intensity in the more oxidized wine, as it displays less of the commonly observed "SO_2 bleaching" effect.

Just as synthetic corks, such as the Nomacorc products just discussed, can have a range of OTR values, an innovative technology has emerged that brings the same ability to control oxygen permeation to screw caps, eliminating the shortcoming of oxygen "starvation" repeatedly found in both the subjective and scientific evaluations mentioned above. The VinPerfect screw cap system is a combination of a conventional screw cap and an inner liner of controlled porosity, as opposed to the nearly impermeable tin/saran liner described above. The system is the invention of veteran winemaker Tim Keller who launched the concept in conjunction with his return to an MBA program at the University of California, Davis Graduate School of Management (GSM). Keller led a team that won the 2008 Big Bang! Business Competition, an annual forum at the GSM that provides startup funds for the top-rated teams. Keller's experience in the wine industry helped him identify a need, and within 2 years, VinPerfect was launched in the Napa Valley, and the SmartCap™ design was introduced in 2011. Well over a hundred wineries have used the system and were able to choose among three liners of different levels of oxygen exposure, similar to the range for natural cork but with more precisely defined values (0.11, 0.21, and 0.49 ppm O_2 per year within a 750 ml bottle). Each liner is a set of four layers: a high-density pharmaceutical grade elastomer (rubber-like material), PET polymer, an aluminum deposition layer, and finally another PET layer providing a food-grade contact surface. PET is an abbreviation for polyethylene terephthalate, a widely used polymer that will be introduced in greater detail as an alternative bottle material in Chapter 15. In 2013, Amcor Limited, the global packaging corporation headquartered in Australia, introduced a similar product line for their Stelvin® aluminum screw caps with four different liners of various OTR values.

After reviewing the many challenges and opportunities for alternative closures, we need to return to cork once again to point out a significant advantage for this traditional material, namely, sustainability, a topic that will be a central focus of our final chapter on alternatives, not to cork but to glass bottles themselves. We briefly noted at the outset of this chapter that the cork tree forests of Portugal and other major cork-supplying countries are sustainable crops. This has significant ecological implications. In the defense of cork, the prolific wine writer Larry Walker has pointed out that the amount of CO_2 required to produce screw cap closures is more than four times that required to produce an equal number of the traditional corks. Furthermore, the Andalusian forests are significant storage sites for CO_2, a fact that is enhanced as a result of regular bark harvesting of the cork trees. Beyond the issue of the carbon footprint, the cork forests are home to a number of endangered species. In an important final defense, Walker points out that the cork industry has substantially improved its quality control efforts and that his experience matches many others in finding that spoilage due to cork taint is now only about 1%.

So we can say, at least for now, that cork closures, like glass bottles, will not go away quickly or quietly. The many positive qualities of the bottle/cork system layered on top of culture and tradition still places it at the heart of wine enjoyment. Finally, we can celebrate this combination by describing a unique technology for opening a wine bottle that utilizes the physical properties of glass in a creative way, *without* removing the cork. Our frequent reference to annealing when discussing glass manufacturing in Chapter 7 speaks to an experience that many will have had, namely, watching a glass product break when subjected to rapid temperature changes. This most commonly occurs when glass is rapidly cooled or "quenched" from a higher to a lower temperature. The annealing steps after glass bottles or stemware are formed acknowledge the need to relieve internal stresses that are produced by the rapid heating and cooling steps that occur during product forming. These stresses are the result of the complex competition among thermal conductivity (heat transfer) and thermal expansion (during heating) or thermal contraction (during cooling) in the heating/cooling steps. The net result of glass fracture due to rapid cooling is appropriately called *thermal shock*, a term introduced in Chapter 6 in relation to an attractive feature of scientific glassware, namely, its resistance to thermal shock damage.

Figures 13.9–13.11 illustrate a long-standing technique for opening a wine bottle without removing the cork by taking advantage of thermal shock. This use of so-called "Port tongs" grew out of a problem with opening bottles of Port wine that were frequently held for long periods of aging in the bottle. In those situations, the very old cork often degraded to the point that the use of a cork-screw would cause the cork to crumble contaminating the wine itself. (Wine consumers among you may have had a similar experience with any wine that has been aged as little as 10 years in the bottle, especially if not cellared at an

Figure 13.9 (a) Port tongs are heated in a burner by (b) Mario Claro, Wine Director at the Six Senses restaurant in the Duoro Valley of Portugal, and then (c) applied to the neck of a vintage Port bottle. (Reproduced with kind permission of Mario Claro.)

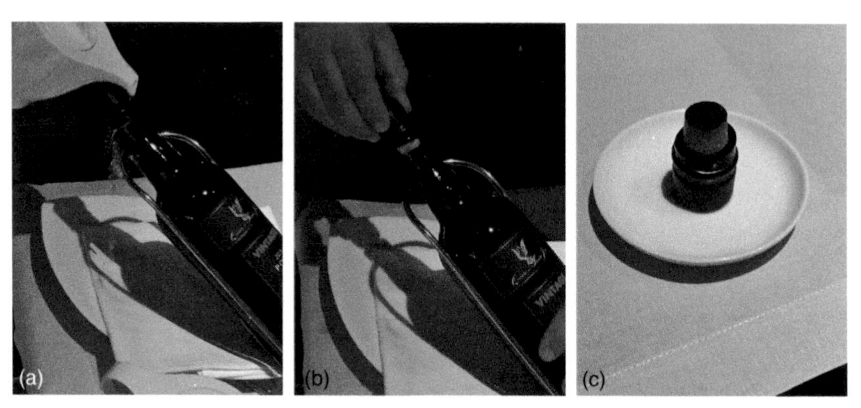

Figure 13.10 (a) After the hot Port tongs are removed from the bottle neck, cold water is poured over the neck creating a clean thermal crack allowing (b) the top of the bottle neck to be removed leaving (c) the embedded cork. (Reproduced with kind permission of Mario Claro.)

optimal temperature [approximately 13°C or 55°F] for its entire life.) The substantial sediment that forms in unfiltered Ports is also a concern, and the use of Port tongs, when skillfully applied, can also minimize the disturbance of that sediment. Figure 13.9 shows how Mario Claro, Wine Director at the restaurant Six Senses in the Duoro Valley of Portugal (the source of grapes used to

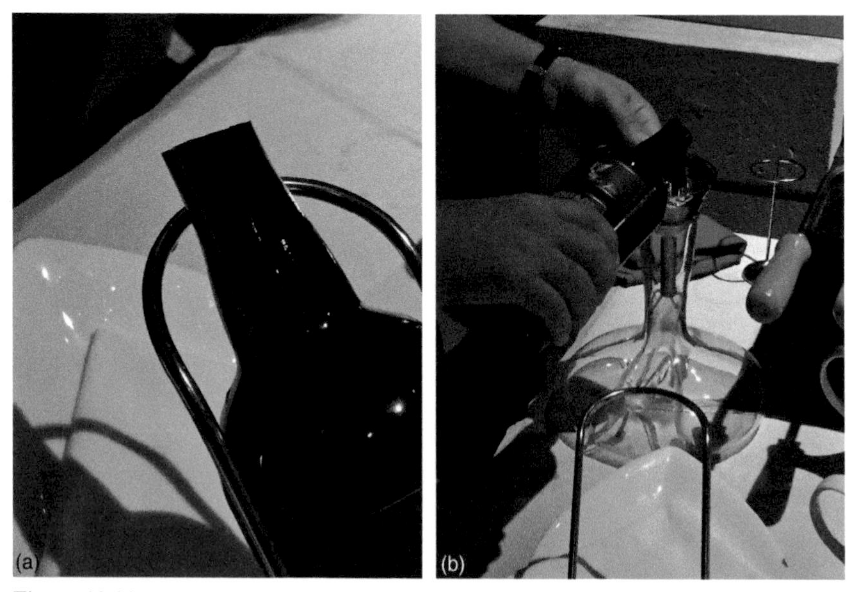

Figure 13.11 (a) The clean break in the neck of the bottle may produce a small glass shard. Because of this fact, along with the likelihood of sediment in the wine itself, (b) the wine is poured through a fine sieve into a decanter before serving. (Reproduced with kind permission of Mario Claro.)

produce Denominação de Origem Controlada [DOC] Port wines), heats the Port tongs over a burner and then applies the cylindrical ring at the end of the tongs to the neck of a bottle of vintage Port. After holding the tongs in place for a few minutes to produce substantial local heating, the tongs are removed, and cold water is poured over the bottle neck creating a remarkably clean break that allows the top of the bottle neck to be removed with the cork intact! (see Figure 13.10). While this technique leads to a clean break in the neck of the bottle (Figure 13.11a), the possibility of a small glass shard and the likelihood of sediment in the wine itself requires that the wine be poured through a fine sieve and decanted, as shown in Figure 13.11b.

This final decanting step leads appropriately into our next chapter discussing the complex relationship between wine and the oxygen in our environment, with oxygen being both wine's friend *and* foe.

BIBLIOGRAPHY

Han, Guomin, Maurizio Ugliano, Bruce Currie, *et al.*, "Influence of Closure, Phenolic Levels, and Microoxygenation on Cabernet Sauvignon Wine Composition after Five Years' Bottle Storage," *Journal of the Science of Food and Agriculture*, **95**, 36–43 (2015).

MacNeil, Karen, *The Wine Bible*, 2nd Edition, Workman, New York (2015).

Shackelford, James, *Introduction to Materials Science for Engineers*, 6th Edition, Pearson, Upper Saddle River, NJ (2005).

Skouroumounis, G.K., M.J. Kwiatkowski, I.L. Francis, *et al.* "The Impact of Closure Type and Storage Conditions on the Composition, Color, and Flavor Properties of a Riesling and a Wooded Chardonnay Wine During Five Years' Storage," *Australian Journal of Grape and Wine Research*, **11**, 369–377 (2005).

Taber, George M., *To Cork or Not to Cork: Tradition, Science, and the Battle for the Wine Bottle*, Scribner, New York (2007).

Walker, Larry,"Cork and Sustainability," *Wines and Vines*, August 2007.

Chapter 14

Perfection through Air – Glass for Aerating and Decanting Wine

"Oxigen (sic), as you well know, is my hero as well as my foe, and being not only strong but inexhaustible in strategies and full of tricks, . . ."

Christian Friedrich Schonbein – Letter to Michael Faraday (December 11, 1860) in *The Letters of Faraday and Schoenbin, 1836–1862*, Williams and Norgate, London, 1899

It seems that most of this book has been devoted to one overarching purpose of glass, namely, keeping wine away from air. Transparent or not, colorful or clear, the myriad optical forms of glass may play a role in the psychology of wine buying, but its essential role as an inert and impervious material has provided centuries of utilitarian service as a great protector of wine from the ravishes of oxidation. On the other hand, we saw in the previous chapter that some "improved" closures that avoid some of the vulnerabilities of cork may prevent the small access to air during the aging process that can give wines aged in cork closures a quality unmatched by some of the newer technologies. Beyond any subtleties of the potential benefits of small amounts of oxygen in the aging of wine, we must now turn to the significant role played by oxygen in optimizing the character of wine once the bottle has been opened. This aeration is especially important for young, tannic reds such as cabernet sauvignon, merlot, nebbiolo, and petite sirah. The not uncommon practice of opening a bottle at the table and "letting it breathe" a while before pouring is an acknowledgment of this need for aeration, but it is a largely empty gesture. Not enough oxygen can reach the

The Glass of Wine: The Science, Technology, and Art of Glassware for Transporting and Enjoying Wine, First Edition. James F. Shackelford and Penelope L. Shackelford.

wine through the narrow neck during a few minutes of exposure to produce a significant effect. What is needed is pouring some or all of the contents of the bottle into a separate container. Once again, it is glass to the rescue.

Decanters are familiar sights in any restaurant with a serious wine list (Figure 14.1). Various designs have evolved over the years, but they serve a common purpose – to provide relatively rapid oxidation of the wine to give a transition from an unacceptably "closed" nature at first opening to a wine that is "open" with optimal character. In fact, the English language is again stretched for the wine vocabulary. Some say that the wine that is "opened" has been "softened" (a rather accurate term for the change wrought in a strongly tannic wine such as a freshly opened bottle of Barolo from Piedmont once described by a favorite wine merchant as roughly equivalent to battery acid).

Although we focus on the relationship between wine and air, the decanter's function goes beyond purely controlling access to oxygen. It can also allow the settling out of sediment commonly found in older red wines. Some pragmatists point out that, in the absence of sediment control, any glass vessel can provide aeration as long as it provides a reasonably large surface area for the wine. Karen MacNeil, author of *The Wine Bible*, was quoted in the *Los Angeles Times* (May 6, 2009) as pointing out that a jelly jar would work just fine. In that same article, Andrew Waterhouse of the Department of Viticulture and Enology at the University of California, Davis, CA was also quoted, pointing out that moving from "closed" to "open" is not merely a story of oxygen. He suggests that the terminology is a bit backward in that the most dramatic chemical process

Figure 14.1 A glass decanter is designed to hold a full bottle of wine and provide substantial aeration, enhanced by a large surface area for the liquid. (Reproduced with kind permission of Riedel Glas Austria.).

occurring immediately upon opening the wine bottle is the release of sulfur compounds that can serve to mask the fresh fruit aromas of the wine.

While beneficial in many cases, decanting should not be a universal practice for red wines. An older pinot noir can be fragile and quickly diminished by excessive exposure to oxygen. The same can be said for older Rioja wines from Spain based on the tempranillo grape and Chiantis based on the sangiovese grape from Tuscany in Italy. Even fresh pinots and others are best not decanted unless some sediment is seen and then just before serving.

As with humans, highly tannic wines (e.g., cabernet sauvignon, nebbiolo, and the Rhone varietals) can become fragile with age (for them, 10 years or more), and all the caution afforded the delicate pinots should be given to these older tannic wines. In this case, decanting should be done within an hour of serving.

When the focus of decanting is the removal of sediment, the bottle should be kept upright for a day or two to allow the sediment to fully settle. Decanting is done until about 5 cm (2 in.) of wine remains; at which point, the sediment should be inspected. Decanting can continue as long as only clear, sediment-free wine results.

On the other hand, decanting is a ritual not limited to reds. The taste of some robust whites can be enhanced by the decanter, but the temperature of the whites can be an even greater factor in controlling the final experience.

The shape of decanters has evolved over the years with a guiding principle being a large surface area for the wine and a vessel that is reasonably easy to pour. Equally important is the aesthetic appeal, as the decanter is after all an integral part of the overall appreciation of a fine wine. A variety of decanter shapes is shown in Figure 14.2.

Figure 14.2 A variety of glass decanter designs demonstrate that, in addition to providing a large surface area for the liquid, the vessel should be easy to pour and aesthetically pleasing. (Reproduced with kind permission of Riedel Glas Austria.).

In the past decade, new technology has entered the field of wine aeration providing both accelerated and miniaturized approaches. The Vinturi® aerator is a noteworthy example with a clever name based on the fact that it draws in a steady flow of air using the Venturi tube principle (Figure 14.3). Ask your favorite fluid mechanician for a quick tutorial or consult Wikipedia should your social circle be limited. In any case, the great sucking sound of the Vinturi gives dramatic evidence that the device is delivering copious amounts of oxygen to a single pour of the wine. In fact, the aeration may be too vigorous for the more delicate varietals such as pinot noir.

Before leaving the Vinturi aerator, we should note that its transparent component is an acrylic polymer, not glass. In any case, the popularity of the Vinturi® has been followed by a number of similar devices. In addition, wine

Figure 14.3 As demonstrated by Wendy Robb, sommelier of the Taco Maria Restaurant in Costa Mesa, California, the Vinturi® aerator provides substantial aeration of a single pour of wine using the principle of the Venturi tube. (Reproduced with kind permission of Daniela Wood.).

glassmakers have joined the effort to deliver air to your wine by producing glasses that "breathe." For example, a German wine glass company, Eisch, marketed a Breathable™ glass in 2005 claiming to provide noticeable aeration in 2–4 min. Court challenges by Riedel glass in 2010 led to a softening of the breathable claim and a change of brand name to Sensis^{plus}. The mechanism by which the Eisch product works is proprietary, and even Karen MacNeil did not seem to know how it worked while she used it in her role as Chair of the Wine Department at the Culinary Institute of America's Greystone campus in the Napa Valley. While the technology is not disclosed, a possible scheme that we can suggest would be to manufacture the glass with residual porosity in the very fine nanometer scale. Such an ultrafine set of connected "bubbles" would not hinder the transparency of the glass. At this point, we can remind ourselves of the rule of thumb from optics raised in Chapter 10, namely, light scattering is most efficient when the scattering centers (e.g., pores) are approximately the same as the light waves (in the wavelength range of 400–700 nm). As noted in Chapter 10, extremely small pores as small as a few nanometers in size can be quite ineffective in scattering light waves, allowing transparency in the presence of porosity. While these nanoscale pores will not hinder transparency, they will allow the easy transport of oxygen molecules.

Whatever the actual technology might be, a word of caution is in order. Rapid aeration can also accelerate overaging. This is especially true in glasses in which "breathing" continues unabated. While the bold cabernet can be brought to its peak within a few minutes, it can become oxidized and "exhausted" before the meal is complete.

As complex as the relationship with oxygen can be for table wines, extra layers of complexity apply to fortified wines such as Port (recall the additional issues raised in Chapter 4 relative to vinifying Port wine). Specifically, Port wine is produced in a variety of ways leading to no single rule for storing and serving (Figure 14.4). In general, Port is cellared at cool temperatures comparable to those for table wines. On the other hand, Port's relationship with air upon opening varies widely depending on the specific type. Unfiltered ones such as a vintage Port require decanting to remove sediment. In fact, the use of Port tongs as described in Chapter 13 (Figures 13.9–13.11) grew out of an effort to minimize the disturbance of that sediment, as well as concerns about a degraded cork. Also, vintage Ports are aged in oak barrels for about 2 years and then transferred to bottles for extended aging, an overall process not so different from table wines. As a result, they need to be consumed within a few days after opening and sooner unless the storage is under an inert atmosphere. On the other hand, tawny Ports are aged for many years in barrel, a process that simultaneously provides some oxidation and evaporation. (The golden-brown "tawny" color is another by-product of this extended barrel aging.) As a result of this substantial exposure to oxygen in barrel, additional oxygen exposure once opened is not as significant, and a tawny Port can be drinkable for 1–2 months with refrigeration.

Figure 14.4 The wines in these bottles of vintage Port (l) and tawny Port (r) have very different relationships with air upon opening, with the tawny Port having already experienced substantial oxygen exposure in its barrel aging and hence remains drinkable for an extended period of time.

And so now, at the end of the chapter devoted to wine and its relationship to air, we can add decanters and other aerating devices designed with glass to the long history of the intimate relationship between this material and the wine it contains, all along a journey that requires the precious nectar to be delivered to us in optimal condition.

BIBLIOGRAPHY

Gray, W. Blake,"Wine Decanting: Give Wines Some Air," *Los Angeles Times*, (2009).

Jordan, Sandra, *The Art of Decanting: Bringing Wine to Life*, Chronicle Books, San Francisco (2006).

MacNeil, Karen, *The Wine Bible*, 2nd Edition, Workman, New York (2015).

Shackelford, James, *Introduction to Materials Science for Engineers*, 8th Edition, Pearson, Upper Saddle River, NJ (2015).

Teichgraeber, Tim,"Riedel Wins 'Breathable Glass' Case," *Decanter*, (2010).

Chapter 15

The Glass of Wine – Now and Forever?

"There is nothing permanent except change."

Heraclitus of Ephesus – as quoted by Diogenes Laertius in *Lives of the Philosophers*

Unlike the culture around beer drinking, glass remains a nearly universally used material for bottle storage and drinking vessels for wine consumption. As we noted at the outset in this book, the image of ordering wine in a Michelin-starred restaurant and having it arrive in something other than a glass bottle and drinking it in something other than elegant stemware (or at least stemless wine glasses) seems inconceivable. We recall from Chapter 13 that breaking from the tradition of the cork closure for the wine bottle has been traumatic for many connoisseurs. There is a similar cultural inertia for using glass, but some pragmatic considerations are creating winds of change, or at least a strong breeze.

The challenge to stemware and the newer stemless variations is not strong. The widespread availability of high-quality but economical wine glasses with the general appearance of those shown in Chapter 12 makes their popularity secure. Of course, the use of plastic imitations for large receptions and ceramic or metal goblets for a traditional context (the medieval feast?) do happen. In 2006, the Hardy Wine Company of Australia introduced an early example of a plastic alternative, with the wine included. Their 187 ml acrylic bottle/"wine glass" set was intended for outdoor events where glass is not permitted due to breakage concerns. Within a year, Al Fresco Wines of England introduced the Tulipak, a biodegradable plastic "glass" with a foil seal that also provided a preset 187 ml serving of wine (with choices of shiraz (syrah), chardonnay, or rosé from Salena

The Glass of Wine: The Science, Technology, and Art of Glassware for Transporting and Enjoying Wine, First Edition. James F. Shackelford and Penelope L. Shackelford.

Figure 15.1 Plasticware comes in a variety of shapes for the convenience of wine appreciation at various informal and outdoor events. Some even look remarkably similar to the traditional wine glass on the right, comparable to the "one-size-fits-all" glass shown in Figure 12.4.

Estate Wines of South Australia). This product was marketed in the United Kingdom and was also intended for simplifying (and making safer with the absence of broken glass) wine service at outdoor events. It was also available at supermarkets across the country. Figure 15.1 shows typical commercially available plasticware vessels that are now widely available for wine consumption at receptions, picnics, and various outdoor events, along with a "one size fits all" glass introduced in Chapter 12.

While the overall challenge to glass drinking vessels may be slight, the relatively large mass of glass bottles in comparison to aluminum, plastic, and cardboard alternatives has led to substantial effort to both design and market containers using these materials, as previously noted in Chapter 11 on glass bottles. The alternatives to glass containers are the ones that have become widely used with other beverages: *aluminum* cans as used for beer and soft drinks, *plastic* bottles as used for soft drinks, as well as *cardboard* (generally coated with another material such as **polyethylene**, the most commonly used polymeric material) as used for milk cartons but, in the case of wine, as wine boxes.

ALUMINUM

Aluminum cans have become commonplace containers for the beer and soft drink industries. An empty aluminum can containing 350 ml (12 oz) of beer or soft drink weighs a mere 15 g in contrast to a glass bottle that could hold the same volume weighing in at about 200 g, a factor of more than ten times! As with the stainless steels introduced in Chapter 4 in conjunction with the revolutionary role of temperature-controlled tanks for winemaking, the chemistry of the metallurgical alloys used in producing aluminum beverage containers are well developed with a few key elements added to the predominantly aluminum composition to give optimal characteristics. In the case of beverage cans, formability is a primary consideration. The contemporary aluminum beverage can is produced in two parts. The bottom cylindrical portion is drawn from a flat plate; hence, the need for extreme deformability. The top closure for the can is then press-formed and attached. The "easy open" tab that prevents environmental waste has been a standard part of can design since 1962, about 5 years after the introduction of aluminum containers and rising complaints about the increasing clutter of discarded opening tabs.

The body of an aluminum can is typically made from an alloy such as aluminum 3004, where the numerical code designates a specific alloy chemistry. In the case of aluminum 3004 and similar alloys in the 3XXX series, the primary alloying elements are manganese (1.0–1.5 wt% for 3004) and magnesium (0.8–1.3 wt% for 3004). While other alloying elements provide specific benefits for the alloy's role as a beverage container (especially ease of forming), the low density of aluminum (2.70 Mg/m^3 compared to 7.87 Mg/m^3 for the predominant element iron in the ferrous alloy used to make the stainless steel winemaking tanks) combined with that ability to form the can with very thin (about 0.10 mm or 100 μm) walls provides the dramatic reduction in container mass that translates into potentially large savings in energy (and thereby money!) in transporting the product to market. A high degree of deformability is not so essential for the top closure of the two-part aluminum can. As a result, a higher strength aluminum alloy is used for that component, such as aluminum alloy 5182 with a lower manganese level (0.20–0.5 wt%) but a significantly higher magnesium level (4.0–5.0 wt%).

It is worth noting that while a major benefit of aluminum for producing beverage cans is its low density (2.70 Mg/m^3), the density of a typical soda-lime-silica glass is even lower (about 2.5 Mg/m^3, depending on the exact glass composition). Unfortunately, the brittle glass can not be manufactured in bottle form with wall thicknesses comparable to that possible for aluminum cans. Chapter 9 attests to this limitation in mechanical behavior. Hence, the much greater volume of glass required for the bottle results in the much greater mass compared to the aluminum can.

In any case, it is interesting to note that, in 1936, the Acampa Winery in Lodi, California marketed a California muscatel in a "tin can" following the emergence of similar cans in the beer market. The Acampa effort, of course, did not become an industry standard, but it does remind us that metal beverage cans were once made of tin, at least partially. As already noted, aluminum alloys were not developed for beverage containers until later, specifically in 1957. Prior to that time, the common metal containers were made of tin-coated steel, with the

Figure 15.2 This pioneering use of an aluminum alloy as an alternative to the traditional glass wine bottle comes with its own straw attached, a dramatic contrast to the association of wine with fine dining, as illustrated in Figure 1.13.

tin providing corrosion resistance. Many people still refer to "tin cans" as the generic term for any metal can, even though aluminum is more likely the primary component.

The famous film director turned major wine producer, Francis Ford Coppola, made one of the most conspicuous entries into this territory in recent years with individual servings of his Sofia Blanc de Blancs sparkling wine (Figure 15.2) in aluminum cans. At about the same time, Mommessin Beaujolais Grande Reserve from France was released in full-sized 750 ml aluminum bottles, and Volute, a wine importer based in San Francisco, began selling single-serving Bordeaux AOC red, white, and rosé wines in 187 ml aluminum bottles. Whether the aluminum container is in a cylindrical can shape or a curvacious bottle and whether it contains beer, a soft drink, or wine, the interior is coated with a polymer due to health concerns about aluminum leaching into the beverage. On the other hand, the standard polymer used for this purpose is bisphenyl A (BPA) that has been banned from its use in baby bottles and sippy cups due to its own health concerns. While criticized as a possible carcinogen and source of other serious health issues, the US Food and Drug Administration (FDA) concluded in 2012 that the level of BPA released from various food containers is safe. Some consumer advocates remain unconvinced.

It is also worth noting that aluminum cans are completely opaque providing the advantage of no light exposure for the wine during storage, as effective and perhaps more so than opaque glass bottles. Aluminum cans also share with metal screw cap closures the ability to prevent any oxygen exposure during storage. As pointed out in Chapter 13 however, this advantage could be a curse, in that a very small amount of oxygen permeation though cork could play a role in the nature of aging of wines stored over long periods of time. With the focus of aluminum cans being used for relatively inexpensive wines, concerns about oxygen deprivation will be a nonissue as long as age worthy cabernets and Barolos are unlikely to be contained in this way.

An overview of aluminum can use versus glass bottles in the last three decades within the US beer industry is given in Figure 15.3. We see that aluminum cans generally represent more than 50% of the beer industry, while the once dominant glass bottles have averaged about one third. In a similar way, aluminum cans represent over 30% of the soft drink industry but are still a nominal fraction of wine sales. While aluminum cans are less than 1% of the wine market at this point, the potential for dramatic savings in transportation costs is driving the reconsideration of this packaging material for at least entry-level wines and possibly even for some more sophisticated offerings.

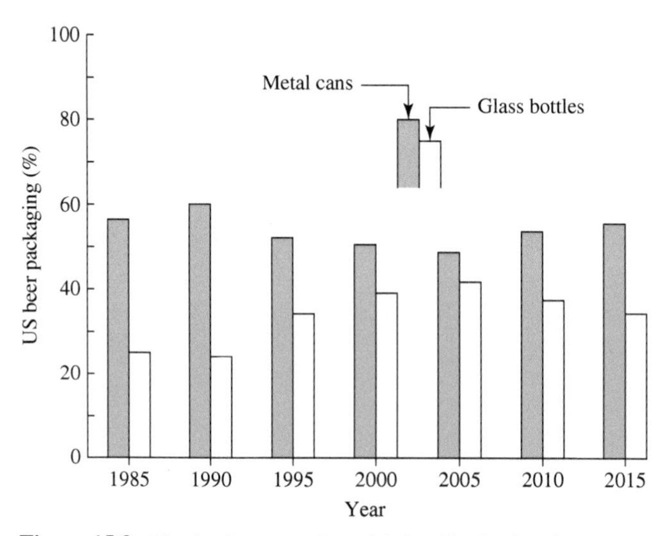

Figure 15.3 The dominant use of metal (primarily aluminum) cans in the US beer industry in the last three decades is shown by this graph. During this time period, glass bottles have represented about one third of the container market on average. Not shown is the roughly 10–20% of the beer market served by draft kegs, as well as the generally less than 1% market share for plastic bottles. (Data from the Beer Institute)

PLASTIC

While aluminum containers are commonplace in the beer and soft drink markets and have significant potential for intrusion into the wine market (with culture and tradition being the primary barriers), plastic bottles have had a limited presence in the beer industry and will likely face similar challenges for wine. The nominal use of plastic as a beer container, as noted in Figure 15.3, is primarily due to its modest level of CO_2 permeability, whereas aluminum and glass have virtually zero. The result is that beer can go "flat" or lose its carbonation after extended storage at even moderate temperatures.

The possibility of compounds in the polymer leaching into beer is an additional concern, and a special problem for beer is that, after bottling, a final, high-temperature water spray is often used as a pasteurization step. The most economical plastics used for soft drink containers can not withstand that process.

Nonetheless, plastics are not without merit. After all, a thin polymer layer protects beer and soft drinks from aluminum leaching, and, of course, their density is even lower than aluminum (about $1.0\,Mg/m^3$ compared to aluminum's $2.7\,Mg/m^3$). So, we should not be surprised that plastic wine bottles have entered the market, albeit with limited acceptance to date. A pioneering example is the Sutter Home wine shown in Figure 15.4 in a bottle composed of **polyethylene**

Figure 15.4 This polyethylene terephthalate (PET) bottle was a pioneering example of wine marketed in a plastic container.

terephthalate (PET), one of the most common polymeric container materials. While PET is a BPA-free material without the controversy associated with aluminum can linings, it still has two significant handicaps. First, there is a small degree of oxygen permeability that limits shelf life from 6 months to 2 years (comparable to the challenge of CO_2 permeation in a plastic beer bottle). Finally, it is worth noting, from the perspective of this book, that the Sutter Home bottle has an ultrathin *glass-like* SiO_x lining to interface with the wine and aid in providing the optimal shelf life. The PET bottle is produced by the Ball Corporation, well known for their glass jars for home canning. The silica lining has the brand name Plasmax, a barrier-coating technology from the German company, KHS GmbH. In 2007, the Saintsbury supermarket chain in Britain introduced wine in PET bottles that weighed 60 g (2 oz) compared to 400 g (14 oz) for their glass counterparts. Clearly, this 85% reduction in weight is a strong incentive for future consideration in the entry-level wine market.

Finally, environmentalists are not limited to a single argument for plastic bottles based on their reduced weight and thus lower energy consumption in shipping product. There is also a substantially lower carbon footprint for manufacturing plastics compared to glasses. The fuel savings for a product that is processed at a few hundred degrees Celsius compared to over 1500 °C for typical glassware is similarly dramatic. The driving force for further consideration of plastic bottles is clear on both fronts – shipping and manufacturing costs.

CARDBOARD

There is perhaps no more humble material to be discussed in this book than cardboard, the substance that daily provides us with milk cartons and an increasingly popular collection of juice packs. We should note that "cardboard" is a highly generic label that covers a range of materials from the beverage containers of our interest to thick corrugated varieties used as heavy shipping boxes. In fact, the term "paperboard" is perhaps a better descriptor, as beverage containers are generally no more than a thick paper. We shall leave to other specialized monographs the interesting history of paper making, the pressing together of moist cellulose pulp, a Chinese invention of the second century AD. Nonetheless, we are reminded of the terminology introduced in Chapter 4 when we discussed wood in order to appreciate that natural material (oak) in its central role in winemaking. Wood is the common source of the cellulose fibers, with the lignin dissolved away in the chemical pulping process.

As a practical matter, beverage containers are not strictly made of cardboard alone. We can easily conjure up an image of wine degrading a purely cardboard carton and slowly oozing out onto a tabletop. Our familiar nonalcoholic examples of cardboard cartons, the milk carton and fruit juice packs, use a polymeric lining and an aluminum/polymer combined lining, respectively. Cardboard-based wine containers follow suit. We shall focus on the two most common examples, namely, the bag-in-a-box and Tetra Pak®.

The bag-in-a-box follows the model of the common milk carton, with one significant difference. Instead of having a polymeric coating on the inner surface, the bag-in-a-box wine container has a flexible polymeric bag that allows the container to shrink in volume, as the wine is poured out. (To be completely accurate about the milk carton, we should note that there is a polymeric coating on both the inside and outside of the cardboard, with polyethylene being the typical polymer of choice.) So, the wine bag is essentially a bladder that collapses along with the shrinking wine volume preventing oxidation and allowing a shelf life of up to several months. A typical example is shown in Figure 15.5. Bag-in-a-box systems often contain as much as 3–5 l of wine, equivalent to 4–7 bottles. Such substantial quantities of *vin ordinaire* can be had for as little as $25 or less.

Tetra Pak is the brand name for the Swiss-based producer of a wide range of food packaging products that originated in Sweden. The increasingly common aluminum foil-lined fruit juice containers are perhaps the best known examples of the technology. Wine often comes in liter-size Tetra Paks, with the example in Figure 15.6 being a half liter (500 ml). One winemaker that uses Tetra Pak has been quoted to say that "A whole Tetra Pak costs less than one cork." A powerful endorsement indeed! The package is composed of a three-material system in a five-layer configuration. As with milk cartons and bag-in-a-box wine containers, a polyethylene layer protects the outside of the cardboard from moisture damage. As one moves into the interior of the container, there are three

Figure 15.5 In this picnic setting, a bag-in-a-box cardboard container encloses a polymer bag that collapses as the contents are emptied, minimizing oxidation and maximizing shelf life up to several months.

Figure 15.6 This Tetra Pak cardboard container is in fact a multilayered packaging system with both an aluminum foil inner liner (having thin polyethylene coatings on both sides of the aluminum) as well as a thin polymeric coating protecting the outside surface of the cardboard.

additional layers – a polyethylene layer between the cardboard and the aluminum foil and, finally, another polyethylene layer to protect the liquid product (wine in our case) from any reaction with the aluminum. As with aluminum cans, the Tetra Pak completely shields the wine from light exposure. While the manufacturer claims that Tetra Pak is recyclable, the intimate bonding of the three components (cardboard, polyethylene, and aluminum) presents a significant challenge with special recycling facilities required and many of the containers ending up in landfills. The recent development of more sustainable systems composed only of cardboard and sugar cane-derived polymers will likely not be practical for wine storage.

Humble or not, cardboard with its ancillary foil materials (plastic and aluminum) shares with aluminum cans and plastic bottles a powerful case for substantial savings and, given some relaxation in cultural traditions, a strong potential for increasing market share in wine storing and shipping. In Australia, such box wines of both types have had more than 50% market share for more than three decades. Tetra Paks for wine have also proven to be quite popular in South America.

Recycling is a persistent issue as we discuss glass containers and their alternatives. We can summarize the relative recycling performance of these three alternative materials in comparison to glass in Table 15.1 with the latest

Table 15.1 Generation, Recovery, and Discard of Container and Packaging Materials in Municipal Solid Waste in 2012[a]

Material	Weight generated (10^9 kg)	Weight recovered (10^9 kg)	Recovered/ generated (%)	Weight discarded (10^9 kg)
Glass	8.51	2.90	34.1	5.61
Aluminum	1.70	0.65	38.0	1.05
Plastic	12.5	1.72	13.8	10.8
Cardboard	34.5	26.1	76.1	8.24

[a]From Municipal Solid Waste Generation, Recycling, and Disposal in the United States: Facts and Figures for 2012, United State Environmental Protection Agency, Washington, D.C., February 2014.

available (2012) data from the US Environmental Protection Agency. All four materials are significant components of the total municipal solid waste in the United States. At the outset, we need to acknowledge that these data include “packaging” as well as containers, skewing the plastic and cardboard numbers away from simply beverage containers. Nonetheless, the table shows that glass and aluminum show comparable efficiencies for recycling. While the overall extent of recycling plastic packaging is small, about 25% of plastic bottles are typically recycled, a value not surprising given the convenient availability of recycling bins. Conversely, the high number for cardboard recycling masks the issues with separating aluminum and polymeric coatings from beverage containers as already mentioned. While the recyclability of glass is competitive with the alternative container materials, the percentages given in Table 15.1 do not tell the whole story. There are technical issues beyond the public’s willingness to recycle, with the data in Table 15.1 driven by both factors. While new glass products can be manufactured with up to 95% recycled glass in the batch, it is important to use recycled glass that is comparable to the desired product. For example, using window glass or scientific glassware with different compositions than standard containers can significantly diminish the quality of the product as well limit the amount of recycled glass allowable in the batch. Color matching is also an important constraint. Recycled colored glass is obviously not acceptable for making a clear glass product. These challenges are occupying both glass engineers and policy makers. Effort on both technical and policy fronts will be necessary to maximize the amount of recycled glass in making new product as well as increasing the public’s use of recycling opportunities, ultimately making glass more competitive against the various alternative materials.

A final, sobering comment about the environmental impact of glass bottles on the overall winemaking process comes from a 2014 article in the journal *Science of the Total Environment*, in which a team of Italian researchers evaluated the impact on global warming of a bottle of the popular white wine vermentino from the large Sardinian producer Sella & Mosca. The research followed the entire viticulture and enology trail from growing and harvesting the grapes to the winemaking process itself and finally the bottling of the wine and its subsequent shipment to consumers in the United States. In monitoring the total energy picture, the study included the energy required to produce the bottles. The dramatic results showed that the bottling/shipping phase represented more than half (56%) of the impact on global warming, while the winemaking portion was roughly one-quarter (27%), and the agricultural production provided the remaining (and smallest) portion (17%). While such estimates are subject to numerous assumptions and results would vary for different wines from different regions going on to their final consumers, the major environmental impact on producing and shipping glass bottles is undeniable. While culture and tradition soundly support the continuing dominance of glass bottles for the wine industry, the pressures from environmental policy and economics can not be expected to diminish with time.

THE "INVERSE BOTTLE"

We can not leave this final discussion of alternative containers for wine without acknowledging a sublimely creative concept from the IDEO design studios in Northern California – not a glass bottle with a cork stopper but *a cork bottle with a glass stopper!* Figure 15.7 shows this clever concept. While it is still in the conceptual phase at this point, this inversion of the materials selection for the centuries old wine container illustrates that the future role for materials in the wine industry is limited only by our imagination. This "imaginary" bottle solves the one limitation of the glass stopper introduced in Chapter 13, namely, the elimination of the need for the small polymeric o-ring shown in Figure 13.8. The idea of a bottle made entirely of cork requires us to reflect on the complex relationship between wine and oxygen, as discussed in Chapter 14. The small amount of oxygen permeation through the traditional wine cork is an important component of the aging process for collectible wines. The substantial increase in the surface area of cork in contact with the wine (a factor of roughly 150 times for a typical bottle/stopper geometry) would affect the aging process but could also be tempered by the use of lining materials of the type discussed for aluminum cans and cardboard containers earlier in this chapter. Of course, there is also the issue of manufacturing the bottle from cork supplies, likely requiring the bonding of cork particulates or laminates reminiscent of the technologies for some synthetic and Champagne corks in Chapter 13. Whether the cork bottle ever moves beyond the conceptual phase, it is a creative, even joyous, reminder of the many innovations we have seen through the millenia covered in this book: the convergence of improved glass transparency and the subsequent clarity of Champagne, the addition of a "modern" material, stainless steel, to the winemaking process and the return to an ancient one, ceramic amphorae, and the

Figure 15.7 This creative inversion of the traditional wine bottle, a glass stopper in a cork bottle, is a concept from the IDEO design studio. While conceptual at this time, it is a provocative approach that builds on many of the principles raised in this book and reminds us that the future of materials selection in the wine industry is limited only by our imagination. (Reproduced with permission of Katie Clark, IDEO.)

ability to make high quality bottles and stemware with machines instead of human powered blowpipes.

EPILOGUE

We close this final chapter looking toward the future and whether the near monopoly of glass as *the* material that defines our relationship with wine will last. In earlier chapters, we looked at the past to understand the history of both wine and glass and how they came to be so intimately intertwined. In a similar way, we needed to understand how both wine and glass are made, journeys through organic chemistry in the first case and physical chemistry in the latter. We also saw that to appreciate the wide spectrum of wines we need to be aware of the shapes of the bottles that contain them and the shape of the vessels in which we consume them. That appreciation required a constant back story, a complex tale of how wines age and especially their relationships to the oxygen that surrounds us and them.

What future lies ahead is impossible to know; true of all futures well beyond our attempt to stare into the crystal ball (glass actually!) about the future of glass in the wine industry. Nonetheless, the inherent qualities of glass, a highly inert surface and the ability to be recycled easily, combined with centuries of tradition as the dominant material for storing, shipping, and consuming wine suggests that it is indeed here to stay. We raise *the glass of wine* to that future!

BIBLIOGRAPHY

Fusi, Alessandra, Riccardo Guidetti, and Graziella Benedetto, "Delving into the Environmental Aspect of a Sardinian White Wine: From *Partial* to Total Life Cycle Assessment," *Science of the Total Environment*, **472**, 989–1000 (2014).

MacNeil, Karen, *The Wine Bible*, 2nd Edition, Workman, New York (2015).

Shackelford, James, *Introduction to Materials Science for Engineers*, 8th Edition, Pearson, Upper Saddle River, NJ (2015).

Taber, George M., *To Cork or Not to Cork: Tradition, Science, and the Battle for the Wine Bottle*, Scribner, New York (2007).

Appendix A

A Primer on Primary Bonding

Throughout this book that spans the topics of organic chemistry (growing grapes and making wine) and inorganic chemistry (making glass to store, transport, and consume that wine), there have been occasional references to the nature of chemical bonding. As such we occasionally made reference to one of the three types of primary chemical bonding. These references were made at the level of a freshman chemistry class. For those for which that class is nonexistent or in the distant past, we provide the following, brief overview of atomic bonding.

Primary bonding involves the transfer or sharing of electrons between adjacent atoms and manifests in one of three types: the ionic, covalent, or metallic bond. To understand atomic bonding, we need to take a look at the structure within individual atoms using a relatively simple planetary model of atomic structure in which **electrons** (the planets) orbit about a **nucleus** (the sun). Figure A.1 is an example for a carbon atom.

In turn, the nucleus is composed of six **protons** and six **neutrons**. Each proton has a positive charge, equal in magnitude but of opposite sign to the negative charge carried by each electron. Note that the neutral carbon atom has six electrons orbiting around the six protons within the nucleus. The neutron is electrically neutral, but contributes to the mass of the element in that each neutron has a mass roughly equivalent to that of a proton. Most of the mass of each element is contained within the nucleus, with the electron having a mass considerably smaller than that of the proton and neutron.

The number of protons defines the **atomic number** for the element (in the case of carbon = 6). This atomic number defines an element's place in the periodic table of the elements, as shown in Figure A.2. The periodicity of the table is based on the elements lining up in chemically similar **groups** (the vertical columns).

The Glass of Wine: The Science, Technology, and Art of Glassware for Transporting and Enjoying Wine, First Edition. James F. Shackelford and Penelope L. Shackelford.

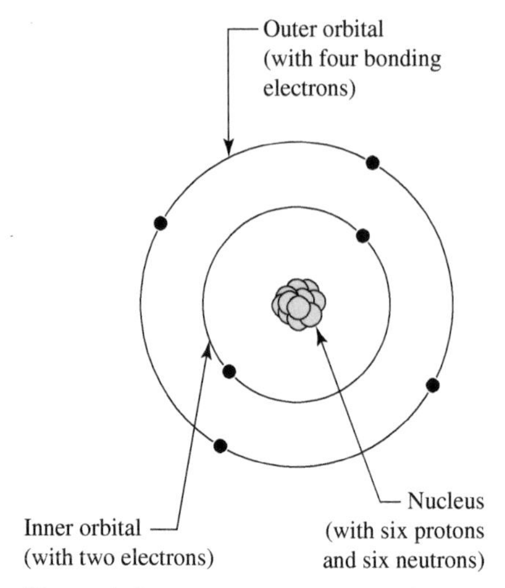

Figure A.1 A schematic illustration of the planetary model of a carbon atom in which six electrons orbit around a nucleus composed of six protons and six neutrons.

It is important to note that the electrons orbiting the nucleus in Figure A.1 are grouped at fixed orbital positions or energy levels, representing the binding energy between the negatively charged electrons and the positively charged nucleus.

The bonding of adjacent atoms to each other involves these orbiting electrons. Strong primary bonds are formed when outer orbital electrons are transferred or shared among adjacent atoms. So, now let us look at the various types of primary bonding, beginning with the ionic bond.

THE IONIC BOND

An ionic bond involves *electron transfer* from one atom to another, as illustrated for sodium and chlorine in Figure A.3. The transfer of an electron *from* the sodium atom to produce the Na^+ ion (or positive *cation*) is favored because it leaves behind a full orbital of eight electrons, which is a more stable configuration. In the same way, the chlorine readily accepts the electron, producing a stable Cl^- ion (or negative *anion*) that also has a full outer **orbital shell** of eight electrons. The ionic bond is the result of the **coulombic attraction** between the ions of opposite charge. This force of attraction is named in honor of the great eighteenth-century French physicist Charles Augustin de Coulomb who demonstrated this attraction of opposite charges using large spheres, not ions.

As noted in Chapter 4, the ionic bond is characteristic of ceramic and glass materials in which oxygen anions (O^{2-}) typically bond to a variety of cations such as Si^{4+}, Al^{3+}, Na^+, and Ca^{2+}.

I A	II A	III B	IV B	V B	VI B	VII B	VIII			I B	II B	III A	IV A	V A	VI A	VII A	0
1 H																	2 He
3 Li	4 Be											5 B	6 C	7 N	8 O	9 F	10 Ne
11 Na	12 Mg											13 Al	14 Si	15 P	16 S	17 Cl	18 Ar
19 K	20 Ca	21 Sc	22 Ti	23 V	24 Cr	25 Mn	26 Fe	27 Co	28 Ni	29 Cu	30 Zn	31 Ga	32 Ge	33 As	34 Se	35 Br	36 Kr
37 Rb	38 Sr	39 Y	40 Zr	41 Nb	42 Mo	43 Tc	44 Ru	45 Rh	46 Pd	47 Ag	48 Cd	49 In	50 Sn	51 Sb	52 Te	53 I	54 Xe
55 Cs	56 Ba	57 La	72 Hf	73 Ta	74 W	75 Re	76 Os	77 Ir	78 Pt	79 Au	80 Hg	81 Tl	82 Pb	83 Bi	84 Po	85 At	86 Rn
87 Fr	88 Ra	89 Ac	104 Rf	105 Db	106 Sg												

58 Ce	59 Pr	60 Nd	61 Pm	62 Sm	63 Eu	64 Gd	65 Tb	66 Dy	67 Ho	68 Er	69 Tm	70 Yb	71 Lu
90 Th	91 Pa	92 U	93 Np	94 Pu	95 Am	96 Cm	97 Bk	98 Cf	99 Es	100 Fm	101 Md	102 No	103 Lw

Figure A.2 The periodic table of the elements.

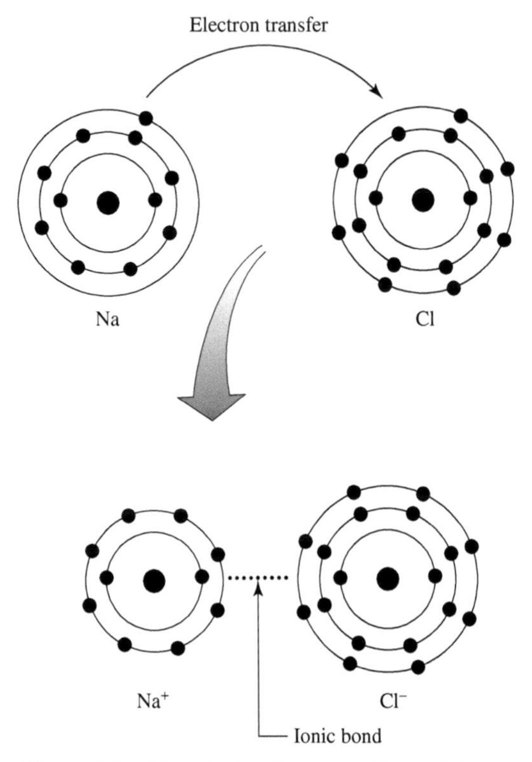

Figure A.3 The ionic bond between sodium and chlorine atoms involves an electron transfer from Na to Cl to create a cation (Na^+) and an anion (Cl^-). The ionic bond is the result of the coulombic attraction between the ions of opposite charge.

THE COVALENT BOND

Unlike the ionic bond, the covalent bond is highly directional. A sodium ion will attract any chlorine ion nearby, but covalent bonding involves the directional sharing of outer shell electrons between specific, adjacent atoms. Those outer orbital electrons that take part in bonding are called **valence electrons**, and the name *covalent* derives from the *cooperative* sharing of those *valence* electrons. Figure A.4 illustrates the covalent bond in a molecule of chlorine gas in which the outer orbital (valence) electrons of each atom are shared between the two atoms.

Before we move on to the third and final category of primary bonding, we must acknowledge that primary chemical bonding is not a black and white issue. Many chemical bonds involve both ionic and covalent characteristics, with no better example than the Si—O bond in the SiO_4 tetrahedron, that is, the building block at the heart of glass structure as shown in Figure 8.1. In fact, the Si—O

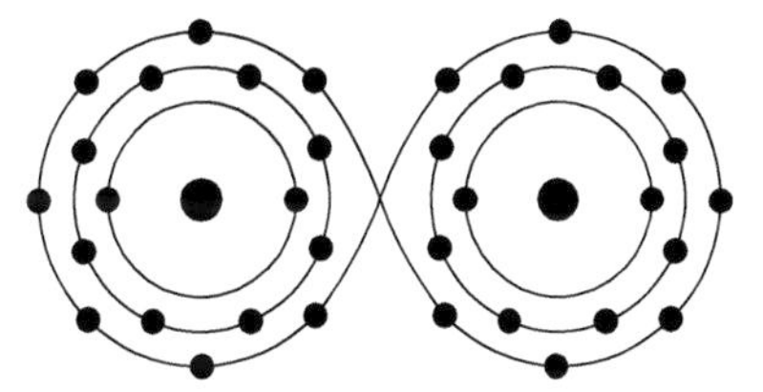

Figure A.4 The covalent bond between two adjacent chlorine atoms produces a Cl_2 molecule. In the molecule, the two atoms share their outer orbital (valence) electrons.

bond is roughly 50% ionic (electron transfer) and 50% covalent (electron sharing) in nature. We should also acknowledge the importance of the *bond angle* concept introduced in regard to Figure 8.3 in relation to covalent bonding, or, as in the case of the Si—O bond, one with partial covalent character. While the $\theta_{\text{Si–O–Si}}$ in Figure 8.3 was seen to vary continuously over a wide range in silicate glasses, the $\theta_{\text{O–Si–O}}$ bond angle in Figure 8.1 is a fixed tetrahedral bond angle = 109.5° and remains essentially the same whether the building block is in a crystalline silicate or in a silicate glass. It is also worth noting that, with carbon residing directly above silicon in the periodic table (Figure A.2), carbon atoms also tend to form many hydrocarbon compounds with a similar fourfold bonding geometry and the same 109.5° tetrahedral angle. While we do not have to delve into the details of the geometry of the complex organic chemistry of winemaking, the fact remains that carbon, at the heart of that chemistry, is a member of the same chemical group (vertical column in the periodic table) as silicon that is at the heart of the chemistry of glass.

One final comment about the covalent bond, as noted in Chapter 4, is that it plays a central role in the nature of polymeric materials. The covalent bonding along a string of carbon atoms in a typical polymeric molecule is relatively strong. On the other hand, the typically lower strength and temperature resistance of common polymers are related to the fact that there is weaker secondary bonding between adjacent chains (Figure A.5). The **secondary bond** is a relatively weak attraction between positive and negative charges within the adjacent polymeric chains but does not involve any transfer or sharing of electrons.

THE METALLIC BOND

The third type of primary bond, the *metallic bond*, is something of a hybrid of the two previous types. We saw that the ionic bond involves electron transfer and is nondirectional in nature, while the covalent bond involves electron sharing and is directional. On the other hand, the metallic bond (like the covalent bond) involves electron sharing and (like the ionic bond) is nondirectional. In this case, the valence electrons are **delocalized**, that is, they have an equal probability of being associated with any of a large number of adjacent atoms.

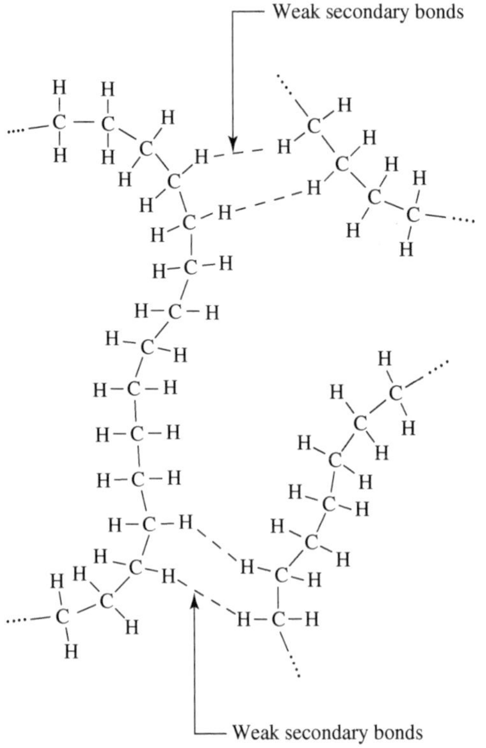

Weak secondary bonds

Weak secondary bonds

Figure A.5 In polymeric materials, weak secondary bonding occurs between adjacent polymeric molecules. Even though strong primary (covalent) bonding exists between carbon atoms along the molecular chain, the weak bonding between molecules contributes to the diminished strength and temperature resistance of polymers in comparison to metallic and ceramic materials.

In typical metals, this delocalization is associated with the entire material, leading to an *electron cloud* or *electron gas* surrounding atomic cores (Figure A.6). The negatively charged mobile "gas" of electrons binds the positively charged atomic cores and is the basis for both the high electrical conductivity of metals and, as pointed out in Chapter 10, their lack of transparency. (The electron gas effectively absorbs light.)

THE CORE ELECTRONS AND COLOR FORMATION

Although we have been able to describe the three types of primary chemical bonding by focusing on the transfer and sharing of outer orbital (valence) electrons, we need to look a bit deeper into the electronic structure of an atom to appreciate the phenomenon of color formation discussed in Chapters 7 and 10.

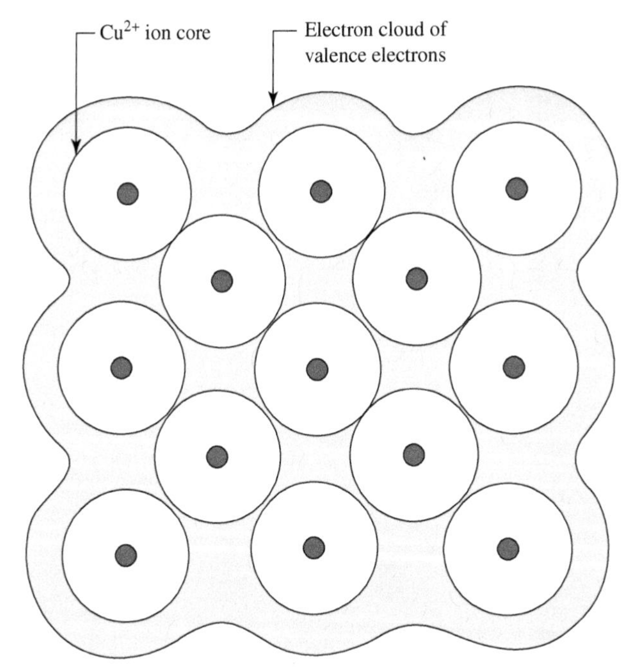

Figure A.6 The metallic bond consists of an electron cloud or gas in which the outer orbital (valence) electrons are highly mobile, providing the glue that holds the atomic cores together. (Each core consists of the nucleus plus the inner orbital electrons.) The case of metallic bonding in copper is illustrated here.

Figure A.7 shows the simple planetary model of electron distributions in a cobalt atom. With an atomic number of 27, cobalt is a considerably heavier element than the carbon atom shown in Figure A.1, and, as a result, has a much larger number of electrons to distribute around the atomic nucleus. Without going into the basis of the ground rules for the population of electron orbitals (quantum mechanics, a topic left for more advanced texts), we can note that the third orbital shell for cobalt contains 15 electrons whereas a "full shell" would have 18. Again without going into quantum details, we can note that this unfilled shell provides mechanisms for electron transitions that are the basis of the absorption of certain photons. (These transitions involve inner shell electrons moving between adjacent orbitals. This movement requires energy that can be supplied by the absorption of the energy of certain photons.) As noted in Chapter 7, cobalt oxide is a classic colorant to produce a distinctive blue glass, and, as noted in Chapters 10 and 11, iron oxide is a widely used colorant for producing green wine bottles. In combination with chromium oxide, iron oxide can also produce a brown or amber color. In general, elements in the periodic table that have incomplete inner orbital shells are termed **transition metals**, with the row from scandium (element 21) to copper (element 29) being an example and

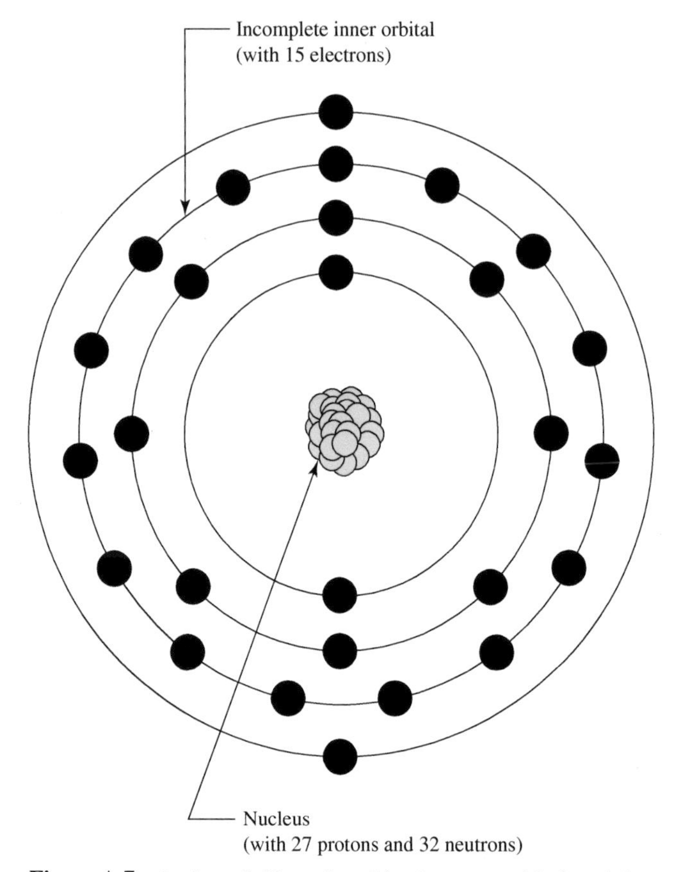

Figure A.7 A schematic illustration of the planetary model of a cobalt atom in which 27 electrons orbit around a nucleus composed of 27 protons and 32 neutrons. It is the incomplete third orbital shell (with 15 electrons compared with a full shell of 18) that gives rise to blue color formation previously shown in Figures 7.7 and 7.8.

including, of course, cobalt (27), iron (26), and chromium (24). Note that the common glass colorants listed in Table 10.1 were all from this row.

BIBLIOGRAPHY

Callister, William and David Rethwisch, *Materials Science and Engineering: An Introduction*, 9th Edition, John Wiley & Sons, Inc., New York (2014).

Hein, Morris, Susan Arena, and Cary Willard, *Foundations of College Chemistry*, 15th Edition, John Wiley & Sons, Inc., New York (2015).

Shackelford, James, *Introduction to Materials Science for Engineers*, 8th Edition, Pearson, Upper Saddle River, NJ (2015).

Appendix B

Glossary

Throughout this book, certain key terms are highlighted in **bold** generally when they are first introduced. These key terms represent the major concepts that are raised in the discussion of the widespread use of glass in the field of wine. Given that we cover a wide spectrum of knowledge, from the physical chemistry of glassmaking to the organic chemistry of winemaking along with the culture of wine appreciation, this glossary is intended to help the reader with a convenient collection of definitions.

alloy A metal composed of more than one element.

anion Negatively charged ion.

annealing A final step in glassmaking for the purpose of relieving stress.

annealing range A narrow band of temperature in which a glass has a viscosity of about 10^{13} poise and internal stresses can be relieved in about 15 min.

atomic number The number of protons in an atom's nucleus. This number defines an element's position in the period table.

bond angle The angle formed by three adjacent, directionally bonded atoms. Monitoring the bond angle, $\theta_{Si–O–Si}$ in Figure 8.3, between adjacent tetrahedra in SiO_2 provides a basic understanding of the difference between crystalline silica (such as quartz) and noncrystalline silica (silica glass).

borosilicate glass Commercial glassware that exhibits excellent chemical durability and, while primarily composed of silica, contains a significant component of B_2O_3.

brittle The lack of deformability.

brittle fracture The failure of a material in the absence of significant ductility.

The Glass of Wine: The Science, Technology, and Art of Glassware for Transporting and Enjoying Wine, First Edition. James F. Shackelford and Penelope L. Shackelford.

°Brix number The index used to indicate the sugar content of grape juice prior to its vinification. This parameter specifically indicates the grams of sugar per 100 g of solution at 20°C and indicates to the winemaker the potential amount of alcohol that will develop in the wine upon fermentation.

bronze An alloy of copper with elements such as tin, aluminum, silicon, and nickel providing both high corrosion resistance and strength.

Bronze Age The period of time from roughly 2000 BC to 1000 BC in which the emergence of metal tools facilitated the development of farming in general and wine grape growing in particular. This time period also represents the foundation of the science of metallurgy.

cation Positively charged ion.

cellulose Compound with the formula $(C_6H_{10}O_5)_n$. Roughly half the composition of wood is cellulose in the form of fibers embedded in a matrix of lignin and hemicellulose.

cement The matrix (usually a calcium aluminosilicate) that encases the aggregate (rock and sand) in concrete.

ceramic A nonmetallic, inorganic engineering material that is distinct from glass in that its atoms are arranged in a regular and repeating fashion, that is, crystalline in nature.

clay A fine-grained soil composed primarily of hydrous aluminosilicate minerals. Amphorae made of fired clay were common storage vessels in the early days of winemaking and are still used by some modern winemakers attempting to return to traditional practices.

color The visual sensation that our eyes can perceive in response to segments across the visible portion of the electromagnetic spectrum with wavelengths between approximately 400 and 700 nm (Figure 10.10).

composite An engineered material composed of a microscopic-scale combination of individual materials from the categories of metals, ceramics (and glasses), and polymers. Wood is an example of a natural composite.

concrete A widely used construction material that is frequently used as a tank for winemaking. It is composed of aggregate (rock and sand) encased in a matrix of calcium aluminosilicate cement.

cork The standard closure for wine bottles made from the bark of the cork tree.

cork taint The production of objectionable tastes and smells in wine due to small amounts of the compound TCA (2,4,6-trichloroanisole) in wine bottle corks.

coulombic attraction The force of attraction between ions of opposite charge.

covalent bond The primary, chemical bond involving electron sharing between atoms.

cristobalite The high-temperature crystallographic form of silica stable just below the melting point, as illustrated in Figure 8.2a.

crystalline Constituent atoms are stacked together in a regular, repeating pattern.

cullet Chunks of broken glass that are recycled into a glass furnace during the manufacturing of new glass products.

delocalized electrons Valence electrons in the metallic bond that are equally probable to be associated with any of the atoms in the metal, corresponding to an electron cloud or gas.

dicotyledons One of the two groups of flowering plants in which there are two (di-)embryonic leaves (cotyledons). The other group is the monocotyledons with one embryonic leaf. Oak, the ubiquitous material choice for wine barrels, is in the dicotyledons group.

diffuse reflection The reflection of light by a rough surface, as illustrated in Figure 10.5.

ductile Capable of being deformed substantially without breaking.

ductility The percent elongation at failure for a material measured in a tensile test, as illustrated in Figure 9.4.

earthenware A low-fired (below 1200°C) ceramic with characteristic porosity. Ancient and modern amphorae for winemaking and transport are examples.

elastic deformation The temporary deformation of a material resulting from the stretching of atomic bonds.

elastic modulus The slope of the stress–strain curve in the elastic region.

electromagnetic spectrum The wide range of radiation frequencies, from the high frequencies of X- and gamma-rays to the low frequencies of radio waves (see Figure 10.2).

electron A negatively charged particle that orbits around a positively charged nucleus.

engineering stress The load on a sample divided by the original (stress-free) area.

fiasco The name of the early, bulbous Italian bottles that contained the Tuscan Chiantis. The characteristic woven straw covering on these bottles served to cushion the fragile vessels from impact damage, as illustrated in Figure 11.2.

The notorious tendency of the bottles to break led to the word "fiasco" becoming synonymous with an ignominious failure.

firing The making of a ceramic or glass product by heating raw materials to a high temperature, typically above 1000°C.

flexural strength (FS) The failure stress of a material, as measured in bending (see Equation 9.2 and Figure 9.7).

gage length The region of minimum cross-sectional area in a specimen examined in a tensile test.

glass The engineered material that is the central focus of this book. It is a noncrystalline solid, meaning that its atoms are arranged in a relatively random way. It has a chemical composition comparable to a ceramic material, where the atoms are arranged in a regular and repeating (or crystalline) way.

glass transition temperature The temperature range, above which a glass is a supercooled liquid, and below which it is a true, rigid solid.

group Chemically similar elements found in a vertical column of the periodic table.

hemicellulose A class of polymers similar to cellulose but with considerably lower molecular weights that, with lignin, forms the matrix in wood microstructures.

index of refraction The parameter that indicates the change in direction of a light beam as it goes from vacuum or air into a glass medium (see Equation 10.3 and Figure 10.3).

International Standardization Organization (ISO) An organization composed of representatives from national standards-setting institutes around the world, including the American National Standards Institute (ANSI) from the United States. The ISO 3591 glass (as described in Table 12.1) is widely used as a standard vessel for sensory evaluation studies of wine.

ion A charged species that results from an electron(s) being added to or removed from a neutral atom.

ionic bond The primary, chemical bond involving electron transfer between atoms.

Iron Age The period of time from roughly 1000 BC to 1 BC in which iron alloys largely replaced bronze for tool (and weapon) making in Europe.

kaolinite A silicate mineral commonly found in clay-based ceramics, such as the ceramic amphorae used in traditional winemaking.

laboratory glassware See *borosilicate glass*.

lignin A class of phenol polymers that, with hemicellulose, forms the matrix in wood microstructures.

matrix The portion of an engineered (or natural) composite material in which a reinforcing, dispersed phase is embedded.

medium-range order A regular arrangement of ions occurring over the range of a few nanometers in an otherwise noncrystalline glass.

melting point The temperature at which a solid transforms to a liquid upon heating.

melting range The temperature range over which the viscosity of a glass is about 100 poise, with a consistency similar to honey.

metal An engineered material having various properties such as ductility and electrical conductivity characteristic of the metallic chemical bond.

metallic bond The primary, chemical bond involving the nondirectional sharing of delocalized electrons.

modifier oxide An oxide such as Na_2O or CaO that, when added to a glass composition, breaks up the linkage of SiO_4 tetrahedra and thereby makes the glass network weaker with the practical result that the glass can be formed at lower (and more economical) temperatures (see Figure 8.5).

neutron A subatomic particle without a net charge that is located in the atomic nucleus, along with positively charged protons.

noncrystalline Atomic arrangement without long-range order.

nucleus The central core of an atom's structure. Electrons orbit around this nucleus.

oak The primary type of wood that is used in making barrels for winemaking. Specifically, four species of oak have proven effective in providing subtle amounts of porosity to allow controlled access of oxygen to the wine during aging, as well as certain phenols that can contribute to the wine's final character.

opaque The inability of a material to transmit an image.

orbital shell A set of electrons within a given orbital.

oxidation The reaction of wine with atmospheric oxygen.

periodic table A graphical arrangement of the elements indicating chemically similar groups (vertical columns) (see Figure A.2).

phenol A class of compounds, the simplest of which is phenol with the formula C_6H_5OH, that serve as precursors to a wide variety of organic compounds. Phenols play an important role in the interaction of wooden barrels with wine.

plastic See *polymer.*

plastic deformation The permanent deformation of a material resulting from the distortion and reformation of atomic bonds.

polyethylene (PE) The most widely used polymeric material, with the chemical formula $(C_2H_4)_n$ where n can have a value of several hundred to several thousand. This formula indicates that *poly*ethylene is a long-chain molecule for which there are several hundred to several thousand ethylene "mers" (each corresponding to a single ethylene molecule with the formula C_2H_4) strung together by covalent bonding between adjacent carbon atoms.

polyethylene terephthalate (PET) One of the most common polymeric materials for beverage containers, now being used for some wine bottles. It has the chemical formula $(C_{10}H_8O_4)_n$.

polymer An organic engineered material composed of long-chain or network molecules.

primary bond One of three types of relatively strong bond between adjacent atoms resulting from the transfer or sharing of outer orbital electrons.

proton A positively charged subatomic particle that is located in the atomic nucleus, along with electrically neutral neutrons.

quartz The room temperature crystallographic form of silica, as illustrated in Figure 3.6.

random network model W.H. Zachariasen's insightful concept that a simple oxide glass can be described as the random linkage of "building blocks" (e.g., the silica tetrahedron).

recycling The reprocessing of relatively inert engineering materials, such as glass bottles. Recycling is a major factor in the competition among materials competing with the traditional glass used in manufacturing wine bottles.

reflection The change in direction of a light beam at an interface, at which the beam returns into the medium from which it originated (see Equation 10.4 and Figure 10.4).

refraction See *index of refraction.*

scientific glassware See *borosilicate glass.*

secondary bond A relatively weak attraction between positive and negative charges within adjacent polymeric chains that does not involve any transfer or sharing of electrons. Secondary bonding accounts for the relatively low strength and temperature resistance of polymers compared to metallic and ceramic/glass materials.

short-range order (SRO) The local "building block" structure of a glass (comparable to the structural unit in a crystal of the same composition). The silica tetrahedron in both crystalline and noncrystalline SiO_2 is an example.

silica Silicon dioxide (SiO_2). Starting as a raw material obtained from common sand deposits, silica becomes the major component of wine bottles, stemware, and other glass products central to the storage, shipping, and consumption of wine.

soda-lime-silica glass A noncrystalline solid composed of sodium, calcium, and silicon oxides. Glass bottles and stemware commonly used for wine storage, shipping, and consumption are in this category. A typical composition is approximately 15 wt% Na_2O, 10 wt% CaO, and 75 wt% SiO_2.

specular reflection The reflection of light by an "average" surface, as illustrated in Figures 10.5 and 10.6.

stainless steel A steel alloy that has substantial resistance to chemical reaction with its environment by the addition of certain alloying elements, especially chromium. Temperature-controlled stainless steel tanks have become widely used in winemaking over the past 50 years.

steel An iron-based alloy that contains up to about 2.0 wt% carbon.

tannin Phenolic biomolecule with distinctive astringency (dry, puckery mouthfeel). Tannins play a significant role in the character of red wine.

tensile strength (TS) The maximum engineering stress measured for a material during a tensile test, as illustrated in Figure 9.4.

tensile test An experimental setup in which a bar of material is pulled in tension to determine its strength and ductility, as illustrated in Figure 9.3.

thermal expansion The tendency for a material to expand upon heating (see Figure 8.7).

thermal shock The fracture (partial or complete) of glass as the result of rapid cooling.

transition metal An element in the periodic table that has an incomplete inner orbital shell, such as the row from element 21 (scandium) to element 29 (copper).

transition metal ion A positively charged ion created from a transition metal atom.

transition metal oxide An oxide compound containing a transition metal ion. These oxides are commonly used to produce colored glass, for example, iron oxide used to produce green wine bottles.

translucent The transmission of a diffuse image.

transparent The transmission of a clear image.

valence electron An outer orbital electron that takes part in atomic bonding.

vanillin Phenolic compound with the formula $C_8H_8O_3$ that is the primary component of the extract of the vanilla bean. This organic compound can be transmitted to wine from oak barrels.

viscosity Liquidlike deformation associated with glasses above their glass transition temperatures.

visible light That portion of the electromagnetic radiation spectrum that can be perceived by the human eye (with wavelengths between 400 and 700 nm) (see Figure 10.10).

wine A fermented beverage generally made from the grape plant (*vitis vinifera*) in which the sugar within the grape is converted to alcohol in the presence of yeasts.

wood A natural composite composed of cellulose fibers embedded in a matrix of lignin and hemicellulose.

working range The temperature range in which glass product shapes are formed (corresponding to a viscosity range between 10^4 and 10^8 poise).

yeast Unicellular microorganism of the fungus kingdom that converts carbohydrates to carbon dioxide and alcohol during fermentation.

yield strength (YS) The strength of a material associated with the upper limit of elastic deformation, as illustrated in Figure 9.4.

Index